REVISE EDEXCEL AS/A LEVEL
Physics

REVISION WORKBOOK

Series Consultant: Harry Smith

Authors: Steve Adams and John Balcombe

A note from the publisher

In order to ensure that this resource offers high-quality support for the associated Pearson qualification, it has been through a review process by the awarding body. This process confirms that this resource fully covers the teaching and learning content of the specification or part of a specification at which it is aimed. It also confirms that it demonstrates an appropriate balance between the development of subject skills, knowledge and understanding, in addition to preparation for assessment.

Endorsement does not cover any guidance on assessment activities or processes (e.g. practice questions or advice on how to answer assessment questions), included in the resource nor does it prescribe any particular approach to the teaching or delivery of a related course.

While the publishers have made every attempt to ensure that advice on the qualification and its assessment is accurate, the official specification and associated assessment guidance materials are the only authoritative source of information and should always be referred to for definitive guidance.

Pearson examiners have not contributed to any sections in this resource relevant to examination papers for which they have responsibility.

Examiners will not use endorsed resources as a source of material for any assessment set by Pearson.

Endorsement of a resource does not mean that the resource is required to achieve this Pearson qualification, nor does it mean that it is the only suitable material available to support the qualification, and any resource lists produced by the awarding body shall include this and other appropriate resources.

For the full range of Pearson revision titles across KS2, KS3, GCSE, Functional Skills, AS/A Level and BTEC visit:
www.pearsonschools.co.uk/revise

Contents

A small bit of small print
Edexcel publishes Sample Assessment Material and the Specification on its website. This is the official content and this book should be used in conjunction with it. The questions in Now try this have been written to help you practise every topic in the book. Remember: the real exam questions may not look like this.

S.I. units

1 Which of the following is not an S.I. base quantity?

☐ **A** mass

☐ **B** length

☒ **C** charge

☐ **D** time

(1 mark)

2 Mechanical work can be expressed in which of the following derived units?

☐ **A** $kg\,m\,s^{-2}$

☐ **B** $kg\,m\,s^{-1}$

☐ **C** $kg^2\,m\,s^{-2}$

☒ **D** $kg\,m^2\,s^{-2}$

> Remember that work = force × displacement and that the units must reflect this.

$E = Mgh$

$kg\ ms^{-2}\ m$
$kg\,m^2\,s^{-2}$

(1 mark)

Guided **3** Link the derived units on the right with their more familiar equivalents on the left. One has already been done for you.

$V = \dfrac{E}{}$

newton	$kg\,m^2\,s^{-3}\,A^{-1}$
volt	$kg\,m^2\,s^{-3}\,A^{-2}$
watt	$kg\,m\,s^{-2}$
ohm	$kg\,m^2\,s^{-3}$

(3 marks)

4 Show that impulse, which is equal to force × time, and momentum, which is mass × velocity, have the same units.

....

.... **(2 marks)**

5 The range, R, of a projectile launched at an angle θ to the horizontal at velocity v in a gravitational field of strength g is given by the equation:

$$R = \frac{v^2}{g}(\sin 2\theta)$$

> $\sin 2\theta$ does not have any units.

Show that the equation is consistent in terms of units.

....

.... **(2 marks)**

Guided **6** The resistance, R, of a wire of length, l, cross-sectional area, A, and resistivity, ρ, is given by:

$$R = \frac{\rho l}{A}$$

(a) Show that an appropriate unit of resistivity is the $\Omega\,m$.

Rearranging the equation gives $\rho = \dfrac{RA}{l}$, and considering the units of the quantities

on the right-hand side gives ....

.... **(2 marks)**

(b) Express $\Omega\,m$ in terms of S.I. base units.

$R = \dfrac{V}{I} = \dfrac{P}{I^2}$, so units are $W\,A^{-2} = $ ....

.... **(2 marks)**

Practical skills

Practical skills

1 A student measures the diameter of a glass rod five times with a micrometer and obtains the results: 5.11 mm, 5.13 mm, 5.10 mm, 5.14 mm, 5.13 mm. The precision of the micrometer is 0.01 mm. The diameter of the rod should be written as:

☐ **A** 5.12 ± 0.01 mm

☐ **B** 5.13 ± 0.01 mm

☐ **C** 5.122 ± 0.02 mm

☐ **D** 5.12 ± 0.02 mm

> Remember that calculating a mean does not improve the precision of the quantity i.e. the number of decimal places.

(1 mark)

2 The thickness of a stack of 50 sheets of paper is measured with callipers and is found to be 6.5 ± 0.1 mm. The thickness of a single sheet of the paper is best written as:

☐ **A** 0.13 ± 0.1 mm

☐ **B** 0.13 ± 0.002 mm

☐ **C** 0.130 ± 0.002 mm

☐ **D** 0.130 ± 0.1 mm

(1 mark)

3 This question is about random and systematic errors in measurements.

(a) Define 'error' in an experimental measurement.

....

.... **(1 mark)**

(b) What is the difference between a random error and a systematic error?

....

....

....

.... **(2 marks)**

Guided

(c) Explain why taking an average of several readings will reduce random error but will not reduce systematic error.

A random error has an equal chance of being positive or negative, so when a

mean is taken ....

....

.... **(2 marks)**

(d) Plotting a graph of one quantity against another can reveal the existence of error in the measurement of those quantities.

(i) How might the graph indicate the presence of a systematic error?

....

.... **(1 mark)**

(ii) How might the graph reveal a significant random error in the measurements?

....

.... **(1 mark)**

Estimation

Guided 1 Estimate the average electrical power required by a small town.

Method: Let's assume the town has 20 000 residents in 5000 homes plus shops, factories and schools, etc. There will be periods of high demand, e.g. evenings, and low demand, e.g. at night. ....

....

....

....

.... **(3 marks)**

2 Estimate the force on your legs if you jump off a 2 m high wall and land on your feet.

....

....

....

....

.... **(3 marks)**

3 It is said that the beam of the Large Hadron Collider (LHC) at CERN has the energy of an express train. If the beam at the LHC has an energy of 360 MJ, is the claim reasonable?

....

....

....

| Start by estimating the mass of the train by comparing it with something more familiar, like a car that has a mass of about 1500 kg. |

....

....

.... **(2 marks)**

4 Estimate how long it would take to fill a swimming pool from your kitchen tap.

....

....

....

....

....

.... **(3 marks)**

SUVAT equations

1 Two cars, each 5.0 m long, are driven at the same constant velocity of 25.0 m s^{-1} in adjacent lanes on a straight road, as shown in the figure below.

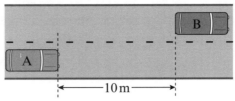

Guided

(a) The driver of car A then decides to overtake car B and accelerates uniformly at 2.0 m s^{-2}. Show that it will take about 6 s until the rear of car A is 15.0 m ahead of the front of car B.

> Don't forget to take the length of each car into account.

If the time is t, B travels $25t$ metres. A (using $s = ut + \frac{1}{2}at^2$) travels $25t + \frac{1}{2} \times 2 \times t^2$ metres. A must travel $10 + 15 + 5 + 5 = 35$ m farther than B.

....

....

.... **(3 marks)**

(b) How far will car B travel during this manoeuvre?

....

.... **(1 mark)**

(c) How far will car A travel during this manoeuvre?

....

.... **(1 mark)**

(d) Determine the final speed of car A after the manoeuvre.

....

....

.... **(2 marks)**

2 A cricket ball is thrown vertically upwards at 30 m s^{-1} ($g = 9.81$ m s^{-2}).

(a) Ignoring air resistance, determine the maximum height that the ball will reach.

....

....

.... **(2 marks)**

Maths skills

(b) Calculate the time at which the ball will return to the point from which it was thrown.

....

....

.... **(2 marks)**

Displacement–time, velocity–time and acceleration–time graphs

1   The **velocity** of a car is described by the graph below.

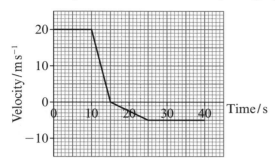

Look carefully at the axes of the *v–t* graph.

 (a)   Using the grid below, sketch a **displacement**–time graph that describes the motion of the car from 0 to 40 s.

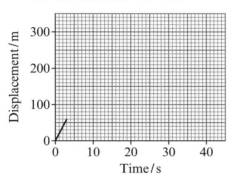

Remember that the area under the line of a *v–t* graph is the displacement.

(4 marks)

 (b)   Using the grid below, sketch an **acceleration**–time graph that describes the motion of the car from 0 to 40 s.

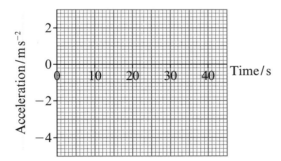

Remember that the gradient of the line of a *v–t* graph is the acceleration.

(4 marks)

 (c)   Calculate the total distance that the car travels from 0 to 40 s.

....

....

.... **(2 marks)**

Scalars and vectors

1   The quantities in the table below are either scalars or vectors. Indicate which is which by a tick in the appropriate column.

Quantity	Vector	Scalar
distance		
momentum		
speed		
pressure		

(4 marks)

Guided

2   (a) Explain why adding together two particular scalar quantities always produces the same result, whereas adding two vectors with the same magnitudes may produce different results.

Scalar quantities are just magnitudes, so adding them up can only produce one

answer, but vectors ....

....

.... **(2 marks)**

Guided

(b) Two forces of magnitude 30 N and 40 N act on a body. Which of the following could equal the magnitude of the resultant force acting on the body? Circle the correct answer(s) and explain.

0 N   10 N   30 N   70 N   80 N   **(2 marks)**

The forces could be in the same direction, when they add up to 70 N, or in opposite directions, when the resultant is 10 N, or they could add up to anything

in between, when they are at an angle to each other. ....

....

(c) Which of the following statements concerning displacement is correct?
    ☐ **A** displacement = velocity × time, and time is a scalar quantity
    ☐ **B** displacement = velocity × time, and time is a vector quantity
    ☐ **C** displacement = speed × time, and speed is a vector quantity
    ☐ **D** displacement = speed × time, and speed is a scalar quantity   **(1 mark)**

(d) A woman drove to her brother's house, which is 60 km due south of her own house, in a time of 1 h. Later that day she drove 60 km due north back to her house, also in 1 h. Which of the following statements is correct?
    ☐ **A** Her average speed was 120 km h^{-1} over the two journeys.
    ☐ **B** Her average velocity was 60 km h^{-1} over the two journeys.
    ☐ **C** Her average speed was 30 km h^{-1} over the two journeys.
    ☐ **D** Her average velocity was 0 km h^{-1} over the two journeys.   **(1 mark)**

3   A student, Alice, argues that temperature is a vector because it can go up or down. Another student, Bikram, disagrees. Who is right and why?

....

....

....

.... **(2 marks)**

Resolution of vectors

1 A wrecking ball is a heavy steel mass on the end of a wire cable. It can be swung into a building that is to be demolished. The ball is pulled back by a horizontal force F and then released. The weight of the ball is 20 000 N.

(a) Add arrows to the diagram to indicate the direction of the weight (W) of the ball and the tension (T) in the wire cable.

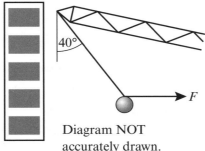

40°

→ F

Diagram NOT
accurately drawn.

(2 marks)

(b) Determine the magnitude of the tension in the wire cable.

> Equate the weight with the vertical component of the tension.

....

....

....   **(3 marks)**

(c) Determine the magnitude of the force F.

....

....   **(2 marks)**

(d) Determine the initial acceleration of the ball in the direction at right angles to the wire cable.

> Consider the component of the weight of the ball at right angles to the cable. You will need g from the data sheet.

....

....

....   **(2 marks)**

2 A box of weight W is pushed up a slope of angle θ against a frictional force F (see figure below). The force up the slope required to move the box at constant speed is:

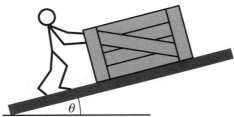

θ

☐ **A** $W + F$

☐ **B** $F + W\sin\theta$

☐ **C** $F + W\cos\theta$

☐ **D** $W + F\sin\theta$

> Considering forces parallel to the slope, there is friction and the component of the weight, $W\sin\theta$

(1 mark)

Adding vectors

1   A duck wants to cross a stream. The duck can paddle at a speed of $0.20\,\text{m s}^{-1}$ relative to the water, which is flowing downstream at $0.10\,\text{m s}^{-1}$.

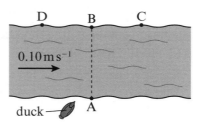

The duck sets off at **A** and paddles toward **B**, which is 2.0 m away, but is carried by the current to **C**.

⟩Guided⟩

(a)   Determine the resultant displacement when it arrives at **C**.

The duck will take 10 s to paddle 2.0 m relative to the water. In this time, it will have been carried 1.0 m downstream, so its displacement will be the vector sum of **AB** and **BC**.

....

.... **(2 marks)**

The duck could have reached **B** by initially heading upstream toward **D**.

(b)   Determine the direction the duck should head relative to the line **AB**.

....

.... **(2 marks)**

(c)   How long will it take for the duck to reach **B** by this method?

....

....

| Remember that the duck's speed is always relative to the water, irrespective of the direction it is moving. |

(2 marks)

2   Two tugboats are manoeuvring a large tanker. At one particular moment, they exert forces of 3000 N and 4500 N, respectively, as shown in the diagram below.

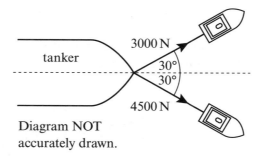

Diagram NOT accurately drawn.

Use a scale diagram to determine the magnitude and direction of the resultant force on the tanker.

(4 marks)

Projectiles

1   A 19th-century cannon could fire a cannonball with a velocity of $480\,\mathrm{m\,s^{-1}}$.
   When mounted high up on a clifftop, this gave it considerable range.
   A ball is fired horizontally from such a cannon, and descends a total distance of
   80 m during its flight.

> You invariably need to
> consider the vertical motion
> first in projectile questions.

Guided

(a) Ignoring air resistance, determine the time of flight of the ball.

Using $s = ut + \frac{1}{2}at^2$ ....

....

.... **(2 marks)**

(b) Determine the horizontal range of the cannon.

....

.... **(2 marks)**

(c) Using the grid below, sketch a graph of the trajectory of the ball as calculated.

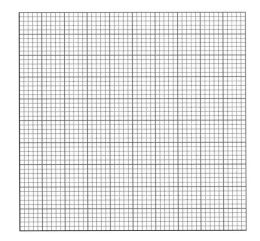

(4 marks)

(d) Add a line to the above graph to indicate the effect of air resistance on the
   trajectory.   **(2 marks)**

Guided

2   When a projectile is launched at an angle θ to the horizontal with initial velocity u,
   its range r is given by:

$$r = \frac{u^2 \sin 2\theta}{g}$$

Explain why this formula predicts a maximum range when θ is equal to 45°.

r will be a maximum when $\sin 2\theta$ is a maximum ....

....

....

.... **(2 marks)**

Free-body force diagrams

1 A mirror that weighs 12.0 N hangs freely from a nail by a length of string attached to the corners of the frame. The mirror has been disturbed from its normal orientation and is hanging as shown in the diagram below.

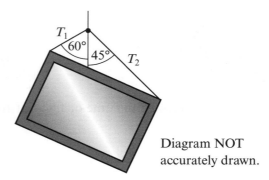

Diagram NOT accurately drawn.

(a) Draw a free-body force diagram that shows the three forces acting on the mirror.

> All the forces should be drawn acting on the centre of mass of the mirror.

(3 marks)

(b) Explain why the upward force on the mirror due to tensions in the string T_1 and T_2 on either side of the nail must have a magnitude of 12.0 N.

....

....

.... **(2 marks)**

⟩Guided⟩ (c) Draw an accurate scale diagram to show that the sum of T_1 and T_2 is 12 N vertically upwards and thus determine the magnitudes of T_1 and T_2.

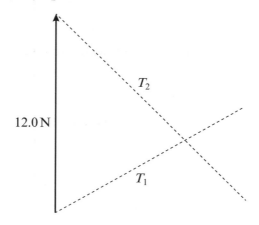

> 'Accurate' means accurate in terms of length and direction.

$T_1 =$ .... $T_2 =$ ....   **(4 marks)**

Newton's first and second laws of motion

1 A car drives along a straight road from **X** to **Y** at constant speed. It then follows the road from **Y** to **Z** without changing speed as the road bends to the left:

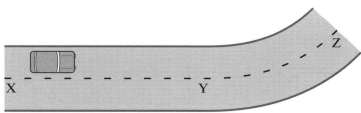

Guided   (a) Explain why the resultant force on the car must be zero as it is driving from **X** to **Y**.

According to Newton's first law of motion, ...

....

.... **(2 marks)**

(b) Explain why there must be a resultant force acting on the car as it drives from **Y** to **Z**.

....

....

.... **(2 marks)**

(c) Circle the diagram that correctly shows the direction of the resultant force on the car at **Y** just as it enters the bend.

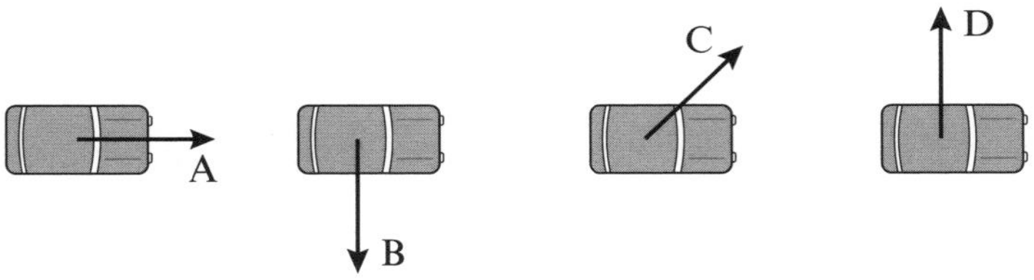

(1 mark)

2 A van of mass 4000 kg is driving at $20 \, \text{m s}^{-1}$ along a straight road. It is exerting a driving force of 4400 N against a drag force of 800 N with other forms of friction being negligible.

(a) Indicate on the above diagram both the driving force and the drag force. **(2 marks)**

(b) Determine the acceleration of the van at this instant.

....

.... **(2 marks)**

(c) If the drag force is proportional to the square of the speed of the van, and the driving force remains constant, what is the van's maximum speed?

....

.... **(2 marks)**

Measuring the acceleration of free fall

1   This question is about measuring the acceleration of free fall using the 'trapdoor' method. The figure below shows the apparatus used. A metal ball is held against electrical contacts. When it is released, a digital timer starts, and when it hits and opens the trapdoor, the timer stops. This enables the time of fall t to be measured for a range of heights h. The results are listed in the table to the right below.

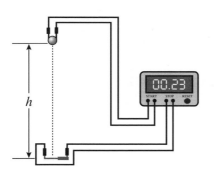

$h \,/\, \text{cm}$	$t \,/\, \text{s}$	$t^2 \,/\, \text{s}^2$
10	0.14	0.020
20	0.20	0.041
30	0.27	0.073
40	0.29	0.082
50	0.32	0.102
60	0.35	0.122
70	0.38	0.143

(a)   Explain why plotting a graph of h against t^2 would be expected to produce a straight line.

> Think of the equation of a straight line $y = mx + c$ and compare it with the equation connecting your variables.

....

....

....

.... **(2 marks)**

Guided

(b)   Describe how you would use the graph to determine the acceleration of free fall of the ball.

The gradient of the graph is equal to $\frac{1}{2}a$, so ....

.... **(2 marks)**

(c)   Plot a graph of h against t^2 on the blank grid here.

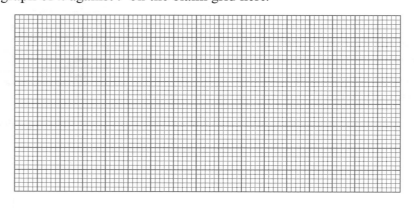

(4 marks)

(d)   Use the graph to determine the acceleration of free fall of the ball.

....

.... **(2 marks)**

Newton's third law of motion

1 An apple sits on a table top. It is in force equilibrium (see figure below).

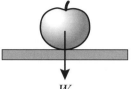

W

> Newton's third law: pairs of forces always act on different objects and are the same type of force.

(a) One force acting on the apple is its weight W. Indicate on the diagram another force acting on the apple that results in force equilibrium.   **(2 marks)**

(b) According to Newton's third law, there must be another force acting as a result of the weight of the apple. What is this force?

....

....   **(1 mark)**

Guided 2 A helicopter can hover by pushing air downwards using its spinning blades.

Explain with reference to Newton's third law how hovering is achieved.

The helicopter's blades push down on the air. Newton's third law of motion says

that for every force there is an equal and opposite force, so ....

....

....   **(2 marks)**

3 A car of mass 1400 kg accelerates at $2.5 \, \text{m s}^{-2}$.

> Remember that forces are vectors and have direction as well as magnitude.

(a) Determine the horizontal force exerted by the road on the car.

....

....

....   **(2 marks)**

(b) What is the horizontal force exerted by the car on the road?

....

....   **(2 marks)**

(c) What is the nature of the above forces (a) and (b)?

....

....   **(2 marks)**

Momentum

1   Which of the following has the greatest linear momentum?

☐ **A**   A 45 kg boy running at 9 m s⁻¹

☐ **B**   A 45 kg cheetah running at 90 km h⁻¹

☐ **C**   A 4.5 g bullet travelling at 900 m s⁻¹

☐ **D**   A 0.045 kg arrow travelling at 90 m s⁻¹

(1 mark)

2   Two balls, P and Q, have equal mass. Ball P, moving at 2 m s⁻¹ to the right, collides with ball Y, which is initially stationary.

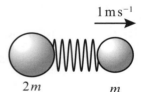

Which of **A**, **B**, **C** and **D** is a possible outcome after they have collided?

	P	Q
A	2 m s⁻¹ to the right	stationary
B	2 m s⁻¹ to the left	stationary
C	stationary	2 m s⁻¹ to the right
D	1 m s⁻¹ to the left	1 m s⁻¹ to the right

(1 mark)

3   Two balls of mass $2m$ and m are held on either side of a horizontal compressed spring. When both balls are released simultaneously, the right-hand ball has an initial velocity of 1 m s⁻¹ to the right.

The initial velocity of the left-hand ball is:

☐ **A**   1 m s⁻¹ to the right

☐ **B**   1 m s⁻¹ to the left

☐ **C**   2 m s⁻¹ to the left

☐ **D**   0.5 m s⁻¹ to the left.

(1 mark)

⟩**Guided**⟩   4   The collision between a golf ball and a club involves a club head of mass 200 g striking a ball of mass 45 g at 31 m s⁻¹. The ball leaves the face of the club at 60 m s⁻¹.

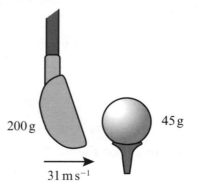

> Just apply conservation of momentum before and after the collision, but remember that momentum and velocity are vectors and have direction as well as magnitude.

Show that the velocity of the club head immediately after the ball leaves contact with it is about 18 m s⁻¹.

Conservation of momentum gives total momentum before impact = total momentum after impact = 0.200 × 31 = 0.200 × v_{club} + 0.045 × 60 ....

.... **(2 marks)**

Moment of a force

1   The figure below shows a painting hanging from a string over a nail. It has been disturbed from its normal horizontal position.

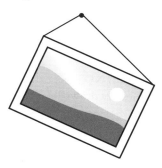

Explain why the painting hangs at this particular angle. Use the diagram to aid your explanation.

....

....

.... **(3 marks)**

2   The figure below shows a railway porter moving heavy goods using a 'sack truck'. The total weight of the sack truck and goods is 2.0 kN.

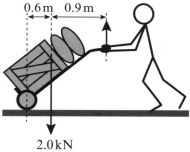

> Consider only vertical forces and horizontal distances. Anything else is irrelevant!

(a)   State the principle of moments.

....

....

.... **(2 marks)**

Guided   (b)   Determine the upward force that the porter must exert to support the load.

Considering moments about the wheel of the sack truck ....

....

.... **(2 marks)**

(c)   Determine the force exerted on the ground by the wheels of the sack truck.

....

....

.... **(2 marks)**

Exam skills 1
Forces, moments and equilibrium

1   This question is about using a winch to raise a horizontal security barrier.

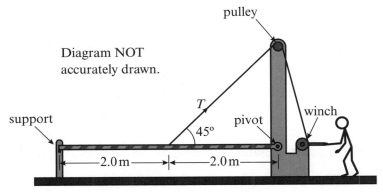

Diagram NOT accurately drawn.

The 4.0 m long barrier weighs 200 N. It is raised by a wire rope attached to its mid-point, which is also its centre of gravity. The wire rope makes an angle of 45° to the barrier when the barrier is in the down position.

(a) The tension in the wire rope is initially zero and the beam rests on a support. Determine the force exerted on the end of the beam by the support.

....

....

....   **(2 marks)**

(b) Explain why the resultant force on the barrier due to the pivot and the support must act vertically upwards.

....

....   **(1 mark)**

(c) The winch is turned so that the barrier just leaves contact with the support but is still horizontal. Determine the tension T in the wire rope under these conditions.

....

....

....   **(3 marks)**

(d) Explain why the force at the pivot cannot act vertically upwards.

....

....   **(1 mark)**

(e) The beam is slowly raised. Explain why the tension in the wire rope decreases as the barrier approaches a vertical position.

....

....

....

....   **(3 marks)**

Work

1   A crate of mass 100 kg is being pushed across a horizontal floor at a steady velocity of 0.5 m s^{-1} with a force of 400 N (see figure below).

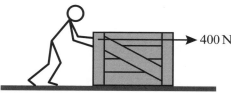

(a) What is the magnitude of the frictional force acting on the box?

.... **(1 mark)**

(b) Indicate the direction of the frictional force on the diagram.   **(1 mark)**

Maths
skills

(c) Calculate the work done in pushing the crate 5.0 m along the floor.

....

....

.... **(3 marks)**

(d) What quantity of heat energy is transferred to the surroundings as a result of the work done above?

.... **(1 mark)**

It is decided that pulling the crate with a rope will be easier (see figure below).

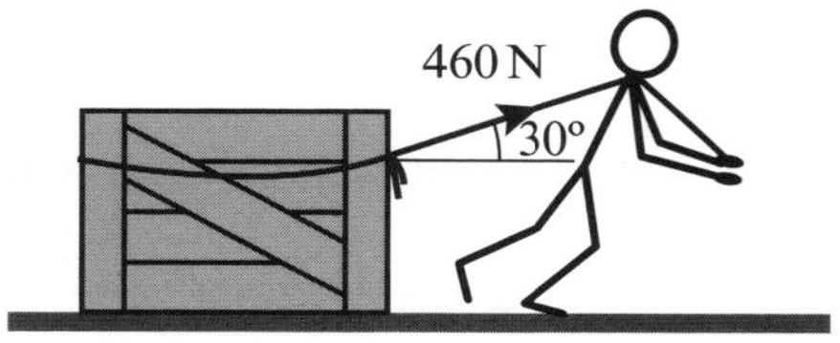

Guided

(e) Why is a larger force now needed to drag the crate along the same floor at the same speed?

Because the pulling force is not horizontal, ...

....

.... **(2 marks)**

(f) How much work is done dragging the crate 5.0 m along the floor?

....

....

.... **(2 marks)**

(g) Compare and account for your answers to (c) and (f) above.

....

....

.... **(2 marks)**

Kinetic energy and gravitational potential energy

1   A car of mass 1800 kg initially travelling at $20 \, \text{m s}^{-1}$ decelerates uniformly to rest over a distance of 40 m in an emergency stop.

(a) Determine the kinetic energy of the car before it starts to decelerate.

.... **(1 mark)**

Guided

(b) Use your result from (a) to determine the mean braking force during deceleration.

The work done by the brakes is equal to the initial kinetic energy of the car =

.... **(2 marks)**

(c) Describe the main energy transfer that takes place while the car is decelerating.

....

.... **(2 marks)**

2   A rubber ball is dropped from a height of 1 m onto a horizontal surface and then bounces several times. Its velocity–time graph is reproduced below. Air resistance can be neglected.

Bounce height is determined by the kinetic energy of the ball as it leaves the surface.

(a) How does the graph show that some kinetic energy is lost as a result of each bounce?

....

.... **(2 marks)**

(b) Why does the bounce height decrease after each bounce?

....

.... **(2 marks)**

(c) What percentage of the ball's kinetic energy is lost as a result of the first bounce?

....

.... **(2 marks)**

Conservation of energy

1 (a) State the principle of conservation of energy.

....

.... **(2 marks)**

   (b) If energy is always conserved, why do we talk about the importance of not 'wasting energy'?

....

....

.... **(2 marks)**

2 This question is about firing an arrow from a bow. In drawing the bow, the bowstring is pulled back by a distance of 0.45 m with an average force of 200 N.

────45 cm────

   (a) How much work is done in drawing the bow?

....

.... **(1 mark)**

Maths skills (b) If 60% of the elastic potential energy stored in the bow becomes kinetic energy of the arrow, determine the kinetic energy of the arrow as it leaves the bow.

....

.... **(1 mark)**

   (c) State what happens to the remainder of the initial energy stored in the bow.

....

....

.... **(2 marks)**

Guided (d) The mass of the arrow is 27 g. Calculate the velocity at which the arrow leaves the bow.

$E_k = \frac{1}{2}mv^2$. Rearranging gives $v = \sqrt{\dfrac{2E_k}{m}} = $ ....

....

.... **(2 marks)**

Work and power

1 At the London Olympics in 2012, Yun-Chol Om of North Korea lifted 168 kg to equal the then world record in his class. In this particular lift, the 'clean and jerk', the weight is first lifted to shoulder height and kept there until the second stage, the jerk, that completes the lift by raising the weight above the head.

(a) The weight was lifted through a total height of 1.50 m. Determine the work done on it during the lift.

....

.... **(1 mark)**

(b) The total time taken for the lift was 5.0 seconds. Determine the average power required to perform the lift.

....

....

.... **(2 marks)**

(c) Explain why his peak output power must have been significantly greater than your answer to (b) above.

....

....

.... **(2 marks)**

⟩**Guided**⟩ 2 A lift in a skyscraper is powered by a 90 kW motor and has a total load capacity of 1800 kg.
Determine the vertical speed that the lift could attain when travelling from the ground floor to the top floor, a total distance of 300 m.

The rate of doing work is the rate of gain of GPE of the lift, i.e. $P = \dfrac{mg\Delta h}{\Delta t}$

....

.... **(2 marks)**

3 A bullet of mass 8.0 g leaves the barrel of a rifle at 700 m s^{-1}. The bullet accelerates uniformly along a barrel of length 0.52 m.

(a) Determine the kinetic energy of the bullet.

....

.... **(1 mark)**

(b) Determine the average power required to accelerate the bullet down the barrel.

....

.... **(2 marks)**

Efficiency

1 A particular car requires a mechanical output power of 12.0 kW when driven at 20 m s^{-1} on a level road.

(a) Calculate the output energy required to complete a 40 km journey at this speed.

....

....

.... **(2 marks)**

Guided

(b) If 1.0 kg of fuel provides 40 MJ of input energy and the overall efficiency of the car is 20%, how much fuel would be used when driving 40 km at 20 m s^{-1}?

If the efficiency is 20%, the input energy is $\dfrac{24.0}{0.20} = 120$ MJ ....

....

.... **(2 marks)**

2 The figure below shows the energy flow associated with a photovoltaic power plant that uses solar panels to convert sunlight into electrical output.

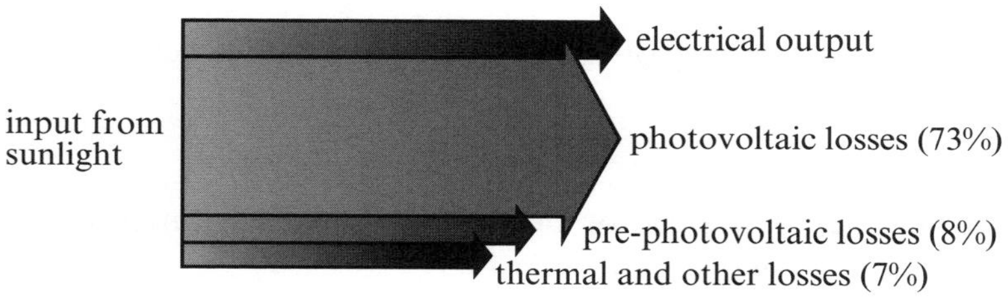

(a) Determine the electrical output as a percentage of the input energy from sunlight.

....

.... **(1 mark)**

(b) The input power from sunlight available for the panels on a particular day is 750 W m^{-2}. What area of panels will be required to produce an electrical output power of 1.8 kW?

....

....

.... **(2 marks)**

Exam skills 2
Forces, energy and motion

1   This question is about driving piles into the ground to reinforce the foundations of a building. The simplest 'drop hammer' method involves lifting a heavy hammer and then simply allowing it to free-fall and strike the pile cap, as shown. The mass of the hammer is 2500 kg. It is raised to a height of 4.0 m above the top of the pile before it is dropped and falls freely. Each time the hammer strikes the pile cap, the pile is driven 0.12 m further into the ground.

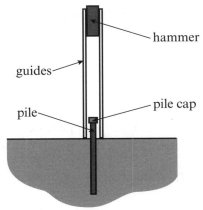

(a) Determine the gravitational potential energy gained by the hammer each time it is lifted above the pile.

....

.... **(2 marks)**

(b) State the kinetic energy of the hammer immediately before it strikes the top of the pile.

.... **(1 mark)**

When the hammer strikes the top of the pile, 80% of its kinetic energy is available to do work in driving the pile into the ground.

(c) Determine the average force between the pile and the ground as it is driven into the ground.

....

....

.... **(3 marks)**

(d) To what form of energy is the kinetic energy of the hammer transferred once the hammer has stopped moving?

.... **(1 mark)**

A more sophisticated pile-driver based on the diesel engine principle can deliver 40 blows per minute to the top of the pile with an energy of 120 kJ per blow.

(e) Determine the average power output of the diesel-driven pile-driver.

....

.... **(2 marks)**

Basic electrical quantities

1 Explain the meaning of the following terms in the context of an electrical circuit:

  (a) Current

    ....

    ....

    .... **(1 mark)**

  (b) Potential difference

    ....

    ....

    .... **(1 mark)**

2 A 9 V battery is connected across a 6 V lamp with a resistor in series with it such that the lamp lights normally (see figure below). The current flowing around the circuit is 0.050 A.

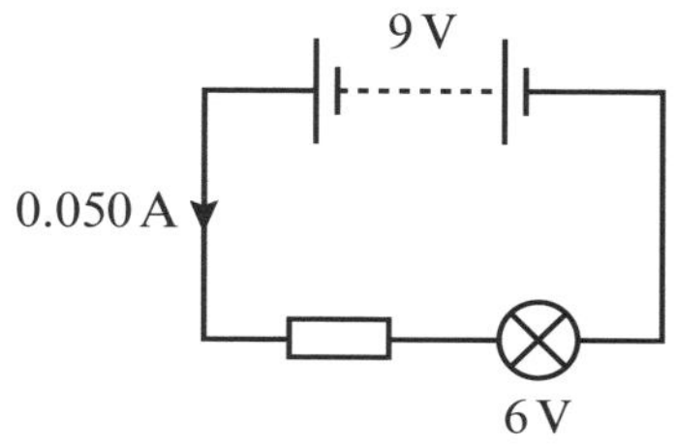

  (a) Determine how much charge flows around the circuit in 100 s.

    ....

    .... **(2 marks)**

  (b) Calculate the electrical energy produced by the battery in 100 s.

    ....

    .... **(2 marks)**

  (c) Determine the electrical energy converted into heat and light by the lamp in 100 s.

    ....

    .... **(1 mark)**

> **Guided**

  (d) Explain **in terms of energy** why the potential difference across the resistor must be 3 V.

    Energy must be conserved, so if the battery is producing 9 J of electrical energy per
    coulomb of charge, and 6 J is converted to heat and light by the lamp per coulomb

    ....

    ....

    .... **(3 marks)**

  (e) Determine the resistance of the lamp.

    ....

    .... **(1 mark)**

Ohm's law

1 Define the quantity **resistance**.

....

....

.... **(1 mark)**

2 The circuit below can be used to measure the electrical characteristics of an electrical component and can also be used to determine its resistance. In this instance, the behaviour of copper sulfate solution is being investigated:

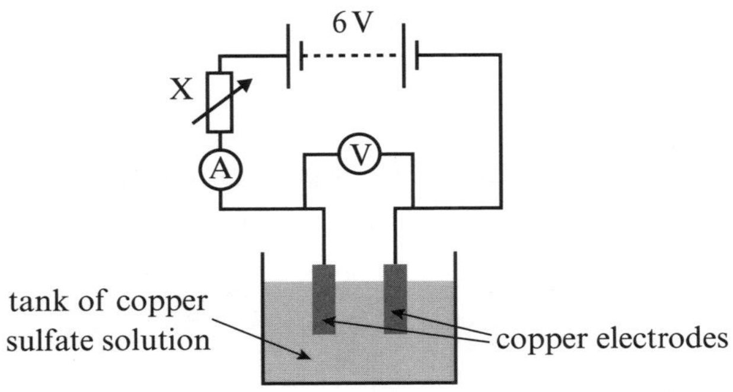

p.d. / V	current / mA
0.00	0
1.13	74
2.21	147
3.30	223
4.38	301
5.51	375

tank of copper sulfate solution — copper electrodes

(a) Name component X and describe its function in this application.

....

.... **(2 marks)**

(b) Name two control variables that must be kept constant during the experiment.

....

.... **(2 marks)**

(c) The data obtained for the current through the copper sulfate solution and the potential difference between the electrodes is listed in the table above. Plot a graph of current against potential difference using the graph below.

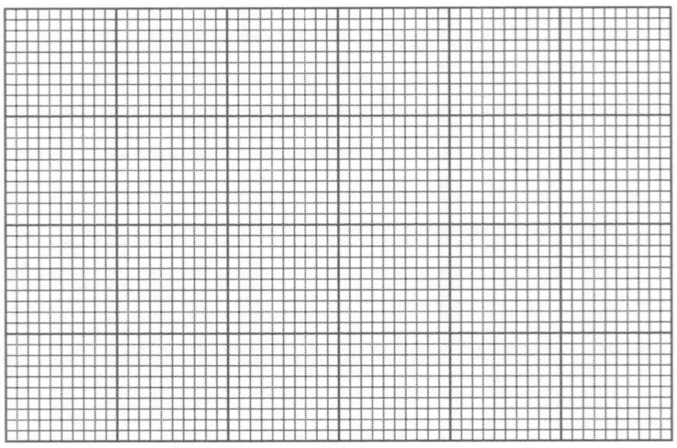

(4 marks)

Guided (d) Explain why it is appropriate to conclude that the copper sulfate solution with copper electrodes is obeying Ohm's law.

The graph is a straight line through the origin so ....

....

.... **(2 marks)**

Conservation laws in electrical circuits

1   The circuit shown in the figure controls model traffic lights. Different combinations of light-emitting diodes require different switches to be operated, leading to different currents in different parts of the circuit.

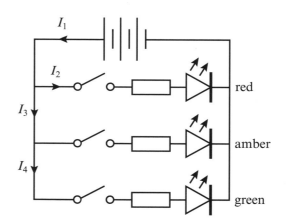

(a)   Complete the following table:

LED combination	I_1 / mA	I_2 / mA	I_3 / mA	I_4 / mA
red		20		
red + amber	45			
green			18	
amber				

(4 marks)

It proves necessary to use three LEDs for each of red, amber and green. The circuit below is for the three red LEDs. The potential difference across each LED is 2.2 V when the current through each is 20 mA. They are powered by a 9.0 V battery.

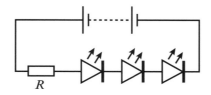

Guided

(b)   Determine the potential difference across R.

The p.d. across R is the e.m.f. of the battery minus the total p.d. across the three

LEDs in series. ....

....

.... (1 mark)

(c)   Determine the resistance of R.

....

.... (2 marks)

Resistors

1   The resistors in the arrangements below are all identical.

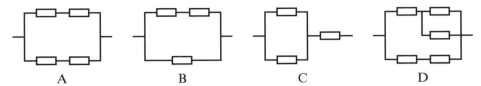

A   B   C   D

(a) Which of A–D has the highest resistance? .... **(1 mark)**

(b) Which of A–D has the lowest resistance? .... **(1 mark)**

2   You have **five** 15 Ω resistors. Draw diagrams to show how you would you produce
    the following resistances using **some** or all of your resistors.

▷**Guided**▷ (a) 20 Ω

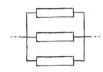

> Three 15 Ω resistors in parallel will provide a useful
> 5 Ω to start you off for (a) as shown here. You only
> have five 15 Ω resistors so you will need to think
> about (b) a bit more. 10 Ω is less than 15 Ω so a
> parallel arrangement is involved again.

(1 mark)

(b) 10 Ω

(1 mark)

3   Determine the resistances of the following combinations of resistors.

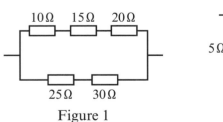

Figure 1   Figure 2

(a) Figure 1:

....

.... **(1 mark)**

(b) Figure 2:

....

.... **(1 mark)**

Resistivity

1 The heating element in a hairdryer consists of a coil of nickel–chrome alloy wire.
The element is connected to 230 V mains and uses a current of 4.35 A.
The resistivity of the nickel–chrome wire is $1.06 \times 10^{-6}\,\Omega\,\text{m}$.

(a) Determine the resistance of the heating element.

....

.... **(1 mark)**

Maths skills

(b) The wire has a diameter of 0.31 mm. Calculate its cross-sectional area in m^2.

....

.... **(1 mark)**

Guided

(c) Determine the length of wire required to construct the heating element.

$R = \dfrac{\rho l}{A}$ can be rearranged to give $l = \dfrac{RA}{\rho} =$ ....

....

.... **(3 marks)**

2 A strain gauge consists of a
network of fine wires made of
an alloy such as constantan
attached to a small piece of
plastic film (see figure).
The strain gauge can be
attached to a metal component,

← strain axis →

for example, and when the latter is deformed, when stressed, the resistance of the
strain gauge will change as the dimensions of the fine wires also change. The change
in resistance indicates how much the component has been deformed.
The strain gauge consists of 12 wires, each 10 mm long by 0.10 mm wide by 5.0 μm
thick. The resistivity of constantan is $4.9 \times 10^{-7}\,\Omega\,\text{m}$.

Guided

(a) Calculate the resistance of the strain gauge.

$5.0\,\mu\text{m} = 5.0 \times 10^{-6}\,\text{m}$ and the twelve wires are in series, so using $R = \dfrac{\rho l}{A}$ gives

....

.... **(3 marks)**

(b) During a particular test, the strain gauge is subjected to a strain of 0.1% along its axis.
If the volume of the wires that make up the strain gauge remains constant, show that the
cross-sectional area of the wires will decrease by about 0.1%.

....

.... **(2 marks)**

Resistivity measurement

1   The following circuit is used to find the resistivity of constantan resistance wire.
A sliding contact allows different lengths l of the wire to be selected.

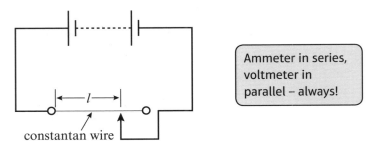

Ammeter in series,
voltmeter in
parallel – always!

constantan wire

(a)   The resistance R of a length l of the wire is to be found by measuring the current
through it and the p.d. across it. Add an ammeter and a voltmeter to the above
circuit diagram to show how these measurements can be made.

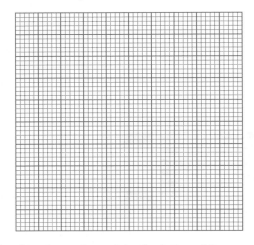

l / m	R / Ω
0.20	1.16
0.40	1.91
0.60	2.95
0.80	3.84
1.00	4.75

(2 marks)

The data is processed to find R and is presented in the table above.

(b)   Plot a graph of resistance against length for the constantan wire using the
grid above.   **(4 marks)**

(c)   Add a line of best fit to the graph.   **(1 mark)**

(d)   Determine the gradient of the line of best fit.

....

....   **(2 marks)**

⟩**Guided**⟩   (e)   The diameter of the wire used was 0.37 mm. Determine its cross-sectional area.

$A = \pi r^2 = \dfrac{\pi D^2}{4} =$ ....

....   **(1 mark)**

(f)   Determine the resistivity of constantan.

....

....   **(2 marks)**

Current equation

1 The number of electrons per unit volume in copper at room temperature is $8.4 \times 10^{28} \, \text{m}^{-3}$.

Guided

(a) Determine the mean drift velocity of electrons in a copper wire of cross-sectional area $1.0 \, \text{mm}^2$ carrying a current of $10 \, \text{A}$.

$I = nqvA$, so $v = \dfrac{I}{nqA} = $ ....

....

.... **(2 marks)**

(b) When a lamp is switched on, it begins to glow almost instantaneously despite the low drift velocity of electrons in a copper wire. Explain how this is possible.

....

....

....

.... **(2 marks)**

(c) Silicon, a semiconductor, has a charge carrier concentration of $8.5 \times 10^{15} \, \text{m}^{-3}$ at room temperature. Discuss, without further calculation, the effect of this difference on the resistivity of silicon compared with that of copper.

....

....

....

....

....

.... **(3 marks)**

2 Resistivity is dependent on temperature in both metals and semiconductors. Compare and account for the effect of temperature on the **resistance** of a filament lamp and a negative temperature coefficient thermistor.

....

....

....

....

....

.... **(3 marks)**

E.m.f. and internal resistance

1 A battery with an electromotive force (e.m.f.) of 6.0 V and internal resistance of 0.2 Ω is connected across a lamp which is **rated** at 6 V, 15 W.

6 V, 15 W

(a) Explain the meaning of 'electromotive force of 6.0 V'.

....

....

.... **(2 marks)**

(b) Explain the meaning of 'internal resistance of 0.2 Ω'.

....

.... **(1 mark)**

(c) Explain why the lamp cannot light with full brightness when connected in this circuit.

....

....

.... **(2 marks)**

2 The circuit is used to determine the internal resistance of a single cell.

The variable resistor allows the current provided by the cell to be varied.

(a) Using the data below, plot a graph of the p.d. across the cell, V, against the current, I, through the cell on the graph paper below.

I / A	V / V
0.10	1.441
0.20	1.433
0.30	1.422
0.39	1.414
0.51	1.406
0.60	1.396

(5 marks)

>Guided>   (b) Determine the gradient of the graph and hence find the internal resistance of the cell.

In this experiment, the gradient of the graph is equal to $-r$ so, as the gradient is

negative, r is a positive quantity. ....

....

.... **(2 marks)**

(c) What is the e.m.f. of the cell?

.... **(1 mark)**

Potential divider circuits

1　Consider the circuits below:

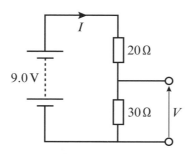

>Guided>　(a)　Calculate the potential difference V across the $30\,\Omega$ resistor.

The potential divider formula $V_1 = \dfrac{VR_1}{R_1 + R_2}$ gives ....

....

.... **(2 marks)**

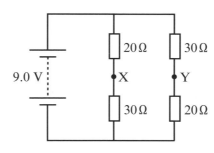

(b)　Determine the p.d. between points X and Y.

....

.... **(2 marks)**

2

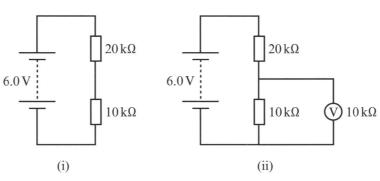

(i)　　　　　　　　　　　　　(ii)

(a)　Determine the p.d. across the $10\,\text{k}\Omega$ resistor in circuit (i) above.

....

.... **(1 mark)**

(b)　A voltmeter that also has a resistance of $10\,\text{k}\Omega$ is connected across the $10\,\text{k}\Omega$ resistor as in circuit (ii) above. What is the reading on the voltmeter?

....

....

.... **(2 marks)**

Had a go ☐ Nearly there ☐ Nailed it! ☐

Exam skills 3
Circuit analysis

1 A battery has electromotive force 12.0 V and an internal resistance of 1.0 Ω.

(a) Explain the term **electromotive force**.

....

....

.... **(2 marks)**

(b) Determine the maximum current that the battery can supply.

....

.... **(1 mark)**

The battery is connected across a lamp as shown.
The p.d. across the lamp is 10.0 V.

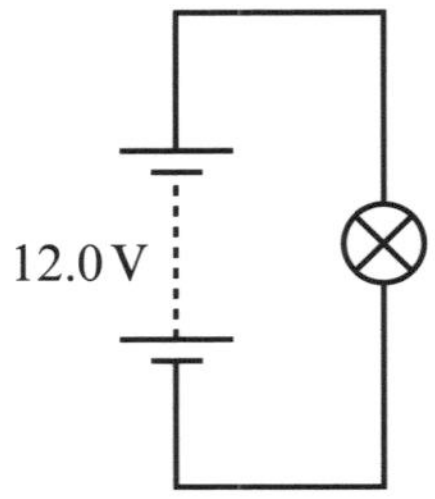

(c) State the p.d. across the battery terminals.

....

.... **(1 mark)**

(d) Account for the difference between the e.m.f. of the battery and the p.d. across its terminals.

....

....

.... **(2 marks)**

(e) Given that lamp A is operating normally, determine the current supplied by the battery.

....

....

.... **(2 marks)**

A second identical lamp is added to the circuit as shown below:

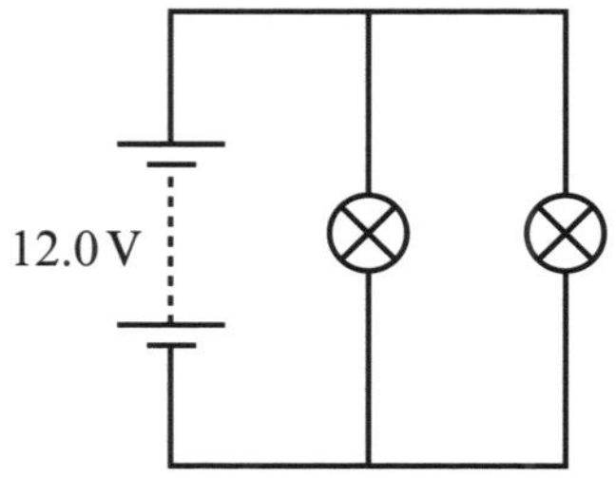

(f) Explain why both lamps light, but with less than normal brightness.

....

....

....

.... **(3 marks)**

Density and flotation

1 A 5.0 cm wooden cube floats in a beaker of water
(density = 1000 kg m^{-3}) as shown in the figure.
The mass of the block is 100 g.

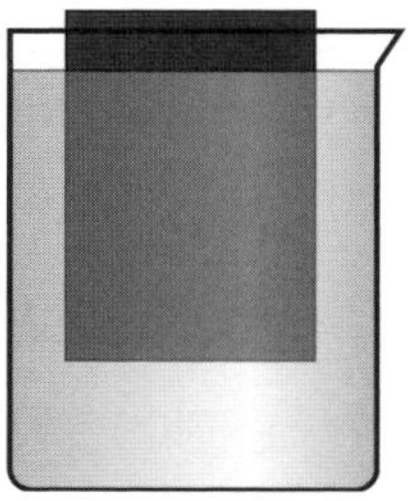

(a) Determine the density of the wooden cube.

....

....

.... **(2 marks)**

⟩Guided⟩ (b) Determine the fraction of the block that is submerged under the water.

The block must displace 100 g ....

.... **(2 marks)**

2 A 1.00 kg mass is suspended from a light thread. It is slowly lowered into a beaker of
water on a top-pan balance. The reading on the balance increases from 500 g to 625 g.

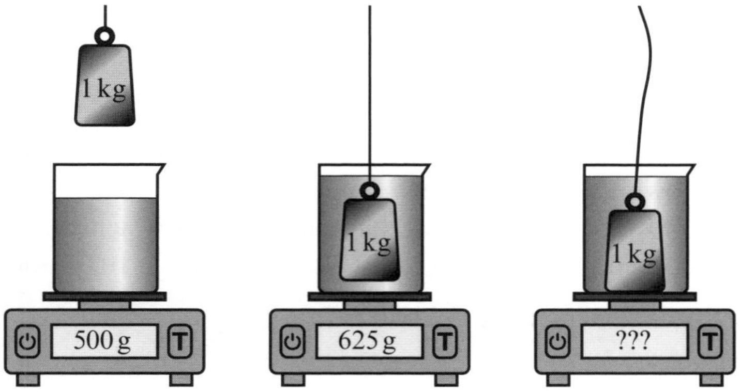

> Think about how
> Newton's third
> law of motion
> might apply to this
> situation.

(a) Determine the tension in the thread before the 1.00 kg
mass enters the water.

> Tension is a force,
> but the balance
> indicates a mass, so
> use $W = mg$ where
> m is in kg and
> $g = 9.81$ N kg^{-1}.

....

.... **(2 marks)**

(b) Determine the tension in the string after the 1.00 kg mass is completely
submerged but not touching the bottom of the beaker.

....

.... **(2 marks)**

(c) Account for the change in the reading on the balance as the mass enters the water.

....

.... **(2 marks)**

The mass is now lowered farther into the beaker until it touches the bottom and the
thread goes slack.

(d) State the reading on the balance.

.... **(2 marks)**

Viscous drag

1 The drag force F on a sphere of radius r moving through a fluid of viscosity η at a speed v is given by the equation for Stokes' law:

$$F = 6\pi\eta rv$$

Guided

  (a) Show that Pa s is an appropriate unit for viscosity.

Rearranging the formula gives $\eta = \dfrac{F}{6\pi rv}$. The 6π can be ignored as it has no

units, so the units of viscosity are ....

.... **(2 marks)**

  (b) The formula for Stokes' law applies to laminar flow only. The ball in the figure below is moving downwards through a fluid exhibiting laminar flow. Add streamlines to indicate the motion of the fluid **relative to the ball**.

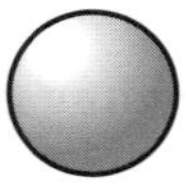

(3 marks)

2 The density of cooking oil is to be measured using the 'falling ball' method. A 1 mm diameter steel ball is dropped down a column of oil and the time it takes to fall 60 cm is measured. This is repeated five times. It is assumed that the ball quickly achieves terminal velocity. The times recorded in seconds were: 10.30, 10.22, 10.25, 10.38 and 10.36.

  (a) Determine the average terminal velocity of the ball in m s^{-1}.

....

....

.... **(2 marks)**

  (b) Draw a labelled free-body diagram to show the forces acting on the ball when it is falling with terminal velocity.

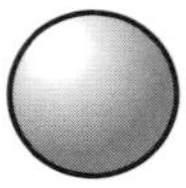

(2 marks)

The viscosity of the oil can be found using the formula: $\eta = \dfrac{2(\rho_b - \rho_o)gr^2}{9v}$

where ρ_b = the density of the ball = 7.85 g cm^{-3} and ρ_o = the density of the oil = 0.92 g cm^{-3}.

  (c) Determine the viscosity of the oil.

....

.... **(2 marks)**

Hooke's law

1 State Hooke's law.

....

....

.... **(2 marks)**

2 The graph shows the behaviour of a spring under increasing loads.

   (a) Determine the load and extension when the spring reaches its limit of proportionality.

....

....

....

.... **(2 marks)**

> **Guided**

   (b) Determine the force constant of the spring.

The force constant k of a spring is defined by the equation $F = kx$, so $k = \dfrac{F}{x}$

or the slope of the straight part of the graph. ....

.... **(2 marks)**

   (c) Determine the elastic potential energy stored in the spring when it is extended by 20 mm.

....

.... **(2 marks)**

3 The graph shows the behaviour of a ductile material under increasing tensile force.

   (a) The material deforms plastically for extensions above 20 mm. How much work is done in increasing the extension from 20 mm to 100 mm?

....

.... **(2 marks)**

   (b) While extending from 20 mm to 100 mm, very little of the work done goes to additional stored elastic potential energy. What happens to the remaining energy?

....

.... **(2 marks)**

Young modulus

1   This question is about the quantities **stress** and **strain** in relation to deforming a material.

(a)   Define **tensile stress** and explain why its units are the same as those of pressure.

....

....

....   **(2 marks)**

(b)   Define **tensile strain** and explain why strain does not have any units.

....

....

....   **(2 marks)**

(c)   Hence explain why the Young modulus has the same units as stress.

....

....

....

....   **(1 mark)**

Guided 2   The Young modulus of copper is 120 GPa. A 3.0 m length of copper wire of diameter 0.50 mm is subjected to a tensile force of 10 N.

(a)   Determine the tensile stress in the wire.

Tensile stress $\sigma = \dfrac{F}{A} =$ ....

....   **(2 marks)**

(b)   Determine the resultant extension of the wire.

$\Delta l = \dfrac{\sigma l}{E} =$ ....

....   **(2 marks)**

(c)   The yield stress of copper is 70 MPa. Determine the maximum tensile force that can be applied to the wire before it starts to deform plastically.

....

....   **(2 marks)**

(d)   The tensile strength of copper is 220 MPa. Why is it not possible to predict the maximum extension of the wire immediately before it fails?

....

....   **(2 marks)**

Exam skills 4 Stress, strain and the Young modulus

1   This question is about measuring the Young modulus of steel using Searle's method.

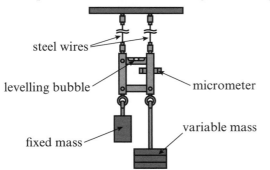

Load / kg	Extension / mm
0.00	0.00
1.00	0.47
2.00	0.99
3.00	1.51
4.00	1.97
5.00	2.46

The apparatus consists of two long, parallel, identical steel wires suspended from a rigid bracket at the upper end and attached to a measuring device at the lower end (see figure above). At the start of the experiment, the levelling bubble is centred using the micrometer. The variable mass can then be increased, which extends the right-hand wire. This then requires the micrometer to be adjusted again by exactly the extension of the wire in order to level the bubble. This enables the extension of the wire to be measured accurately as the load is increased.

(a)   Explain how using two parallel wires can reduce the effect of temperature changes on the accuracy of the experiment.

....

....

.... **(2 marks)**

(b)   Plot a graph of load against extension using the data provided in the table on the right above.

(4 marks)

(c)   Determine the gradient of the graph.

....

.... **(2 marks)**

The wire has a diameter of 0.60 mm and a length of 3.00 m.

(d)   Determine the cross–sectional area of the wire in m².

....

.... **(2 marks)**

(e)   Show that the Young modulus of steel is about 210 GPa.

....

....

....

.... **(2 marks)**

Waves

Guided 1 A radio station transmits radio waves with a frequency of 96.1 MHz. A simple receiving aerial on a radio consists of a telescopic metal tube with a length equal to one quarter of the wavelength of the radio waves it is designed to receive. Given that radio waves travel at 3.00×10^8 m s⁻¹, determine the length of aerial required to receive a 96.1 MHz signal.

Rearranging $v = f\lambda$ gives $\lambda = \dfrac{v}{f}$ .... $\dfrac{3 \times 10^8}{96.1 \times 10^6} = 3.12$ $\times 0.25$

.... 0.78m .... **(2 marks)**

2 Ultrasound waves with a wavelength of 0.44 mm are used for medical imaging purposes. The speed of ultrasound waves in soft tissue is typically around 1540 m s⁻¹.

(a) Determine the frequency of such ultrasound waves.

.... $v = f\lambda$ .... $\dfrac{v}{\lambda} = f$ .... $\dfrac{1540}{0.44 \times 10^3} =$

.... 3.5M.H₂ .... **(2 marks)**

(b) What is the advantage of using ultrasound waves with a very short wavelength?

.... They will diffract more and ....

.... Prode a clearer Image ....

— Redue diffraction

— Increase resolution.

> Think about the effect of wavelength on diffraction and the ability to observe fine detail.

.... **(2 marks)**

3 The graphs below show how the particle displacement y varies with position x along a wave and time t, respectively:

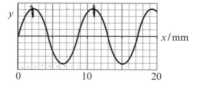
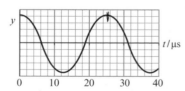

(a) Determine the wavelength of the wave.

.... 9mm ....

.... **(1 mark)**

(b) Determine the frequency of the wave.

.... 25µs .... $\dfrac{1}{25 \times 10^{-6}} = 40000$Hz ....

.... **(2 marks)**

(c) Determine the speed of the wave.

.... $v = f\lambda$ .... $40000 \times 9 \times 10^{-3} = 360$ms⁻¹

.... **(1 mark)**

Longitudinal and transverse waves

1   The figure below represents a 'snapshot' of a progressive wave.

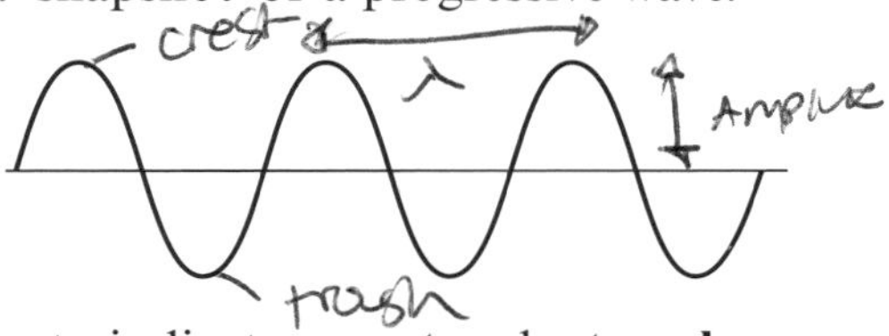

crest λ *Amplitude*

trough

   (a)   Add labels to the diagram to indicate a **crest** and a **trough**.   **(2 marks)**

   (b)   Identify and clearly label the **amplitude** of the wave and the **wavelength** of the wave.   **(2 marks)**

Guided

2   Describe the difference between a **transverse** wave and a **longitudinal** wave. You should use diagrams to aid your explanation.

oscillation

oscillatn

energy

energy

In the case of transverse waves, the particles carrying the wave oscillate at right

angles to the direction of propagation of the wave or the direction of energy

transfer. For longitudinal waves .....the particle oscillate parallel

.....to the direction of energy transfer.....

.....

.....

.....

.....   **(4 marks)**

3   Which of the following waves is longitudinal?
   ☐ **A** ultraviolet
   ☐ **B** X rays
   ☒ **C** ultrasound
   ☒ **D** ripples on water   **(1 mark)**

4   Transverse waves differ from longitudinal waves in that they can be:
   ☐ **A** diffracted
   ☒ **B** polarised
   ☐ **C** refracted
   ☐ **D** reflected   **(1 mark)**

5   When a sound wave travels through air, a point of maximum particle displacement is also a point of:
   ☒ **A** maximum pressure
   ☐ **B** zero pressure
   ☐ **C** minimum pressure
   ☐ **D** maximum wave speed   **(1 mark)**

Standing waves

1 The apparatus depicted is to be used to investigate microwave standing waves. The microwave probe can be moved in the space between the microwave transmitter and the metal sheet that acts as a reflector.

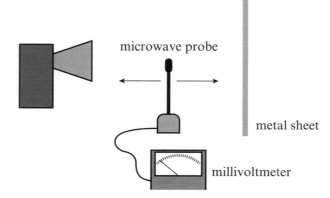

microwave probe

metal sheet

millivoltmeter

Guided (a) Explain how microwave standing waves are produced in the space between the microwave transmitter and the metal sheet.

Probe detects a direct wave & a reflected wave giving two waves of similar

amplitude but travelling in opposite directions .... The tho waves

.... Moving in opposite directions will superpose

and here the trough meets a trough you will
crest   cest

get an antinode. And when a trough meets on   **(4 marks)**
       crest you get on antinode.

in phase =
Antinode

Antphase
= node.

(b) The reading on the millivoltmeter V varies periodically as the probe is moved from the metal sheet toward the transmitter by a distance x as illustrated by the graph below:

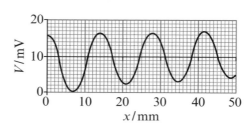

Account for the variation in the signal picked up by the probe.

−close to the sheet here is a high voltage because

here is an antinode

.... **(3 marks)**

(c) Determine the wavelength of the microwaves used in the experiment.

    $\lambda = 14$ $14 \times 2 = 28$ mm

.... **(2 marks)**

Phase and phase difference

1   Describe, in terms of phase, the movement of particles:

>Guided>   (a)   between two adjacent nodes in a standing wave.

The particles between two adjacent nodes in a standing wave always ....

.... **(1 mark)**

(b)   at two adjacent antinodes in a standing wave.

.... **(1 mark)**

2   Explain the meaning of the term **wavefront**.

....

.... **(1 mark)**

3   The figure represents
two progressive waves,
A and B.

y-displacement

x-displacement

(a)   Determine the phase difference in **radians** between wave A and wave B.

....

.... **(2 marks)**

(b)   State the phase difference between A and B in **degrees**.

.... **(1 mark)**

4   The speed of sound in air is $340\,\text{m s}^{-1}$. Two loudspeakers A and B emit sounds of frequency $850\,\text{Hz}$ and are in phase with each other.

(a)   Determine the phase difference between the sound arriving at a point P that is $3.00\,\text{m}$ from A and $3.20\,\text{m}$ from B.

....

....

.... **(2 marks)**

(b)   The connections to loudspeaker A are reversed so that the sound from A is now in antiphase with that from B. What effect will this have on the phase difference between the sounds from A and B arriving at P?

....

.... **(1 mark)**

41

Superposition and interference

1   Two waves in antiphase will produce zero resultant when superposed only if they have the same:

☐ A wavelength   ☐ B frequency   ☐ C amplitude   ☐ D velocity   **(1 mark)**

2   When two waves of frequency f superpose, the resultant has a frequency of:

☐ A f ☐ B $\sqrt{2}f$ ☐ C $2f$ ☐ D $4f$ **(1 mark)**

3   To be coherent, two waves must have the same:

☐ A amplitude   ☐ B phase   ☐ C frequency   ☐ D polarisation   **(1 mark)**

4   The figure below is a displacement–time graph of two waves, X and Y.
Add to the diagram the resultant displacement–time graph that would result from the superposition of waves X and Y.

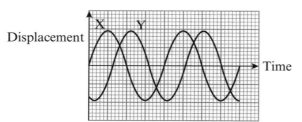

Displacement ———→ Time

> The resultant is found by adding the displacements of the two waves. It is easy to work this out when one of the waves has zero displacement.

(3 marks)

5   A microwave transmitter emits microwaves of wavelength 2.8 cm. The microwaves arrive at a receiver by travelling directly along the line **AB** and also along the path **ACB** after reflecting off a metal plate placed at the centre point **C** (see figure below).

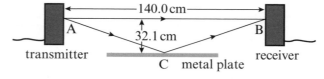

transmitter   C   metal plate   receiver

>Guided>   (a) Determine the path length ACB in cm.

Using Pythagoras' theorem: **ACB** = $2 \times \sqrt{70.0^2 + 32.1^2}$ ....

.... **(2 marks)**

(b) Determine the path difference between **ACB** and **AB**.

.... **(1 mark)**

(c) Determine the resultant phase difference between microwaves travelling along path ACB and those travelling directly from A to B.

....

.... **(2 marks)**

(d) State the type of interference at B associated with the phase difference in (c) above.

.... **(1 mark)**

Velocity of transverse waves on strings

1   Which one of the following changes to the properties of a metal wire under tension, if made alone, will **decrease** the velocity of transverse waves along the wire?

☐ **A** increasing the length of the wire

☐ **B** decreasing the length of the wire

☐ **C** increasing the density of the metal

☐ **D** decreasing the diameter of the wire   **(1 mark)**

2   Which of the following wires A–D, with the same thickness and made of the same material, will vibrate with the **highest** frequency when plucked in the middle?

	Tension	Length
A	T	l
B	$2T$	l
C	T	$2l$
D	$2T$	$2l$

(1 mark)

3   In an experiment to measure the velocity of transverse waves along a wire under tension, the apparatus shown below was used. The wire can be forced to vibrate at different fundamental frequencies f as the tension is varied. The length l remains constant. The data obtained are shown in the table on the right.

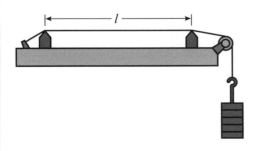

T / N	f / Hz	f^2 / Hz^2
10.0	39.6	1570
15.0	48.2	2320
20.0	56.0	3140
25.0	62.8	3940
30.0	68.2	....
35.0	74.4	....

(a)   Fill in the two missing values in the data table.   **(1 mark)**

(b)   Plot a graph of f^2 against T on the blank grid below using the data provided.

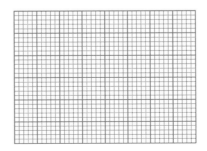

(4 marks)

(c)   How does your graph show that f^2 is directly proportional to T?

....

....   **(2 marks)**

The behaviour of waves at an interface

1 A ray of light is incident at an angle of 50° at an air–glass interface. The refractive index of the glass is 1.49.

Tick **any** of the following statements that apply to the ray at the air–glass interface.

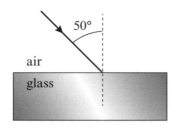

A	The ray will be refracted away from the normal.	
B	The ray will be partially reflected.	
C	The angle of refraction will be about 30°.	
D	The ray will be totally internally reflected.	

(2 marks)

2 An ultrasonic 'tape measure' can determine the distance to a remote object by sending out a stream of ultrasound pulses and measuring the time taken for the pulses to return to the device. Each pulse consists of 10 complete waves of 40 kHz ultrasound. 20 pulses are sent out every second.

> **Guided**

(a) Determine the duration of each pulse.

The time for one cycle of ultrasound is found using $T = \dfrac{1}{f}$. A pulse contains

10 cycles so the duration is ....

.... **(1 mark)**

(b) Given that the speed of ultrasound is 340 m s⁻¹, determine the total distance that one pulse could travel to a target and back before the next pulse is transmitted.

....

....

.... **(2 marks)**

(c) Determine the wavelength of the ultrasound waves.

.... **(1 mark)**

(d) Suggest what the maximum distance that the device could measure might be. Explain your reasoning.

....

....

.... **(2 marks)**

Refraction of light and intensity of radiation

1 A green laser pointer produces light with a wavelength of 532 nm in air.

(a) The speed of light in air is 3.00×10^8 m s^{-1}. Determine the frequency of the green light.

....

.... **(2 marks)**

(b) When light enters water from air, it is slowed down by a factor of 1.33, the refractive index of water. Complete the following table for light from the laser pointer.

	In air	In water
Speed / m s^{-1}	3.00×10^8	2.26×10^8
Wavelength / nm	532	....
Frequency / Hz	....	....

(3 marks)

Guided

2 A lamp emits light uniformly in every direction. The power of the lamp is 100 W, and it can be treated as a point source of light. If the efficiency of the lamp is 7.5%, what is the intensity of the light produced by the lamp measured at a distance of 2.0 m from the lamp?

The area of a sphere of radius r is given by $A = 4\pi r^2$ and intensity is $\dfrac{power}{area}$

....

....

.... **(3 marks)**

3 A ray of light enters a glass tank of water at an angle of 40° (see figure).

40°

Diagram NOT accurately drawn.

air   glass   water
$n = 1.00$ $n = 1.52$ $n = 1.33$

Guided

(a) Determine the angle of refraction at the air–glass boundary.

As $n\sin\theta$ = constant, at the air–glass boundary we can say that $n_{air}\sin 40° = n_{glass}\sin\theta_g$

....

....

.... **(2 marks)**

(b) Determine the angle at which the ray enters the water.

....

.... **(2 marks)**

Had a go ☐ Nearly there ☐ Nailed it! ☐

Total internal reflection

1 (a) Define **refractive index**.

....

.... **(1 mark)**

(b) Define (with an expression) and explain the term **critical angle**.

....

....

....

.... **(3 marks)**

(c) A 45° prism is made from acrylic plastic that has a refractive index equal to 1.49. A ray is incident, as shown below:

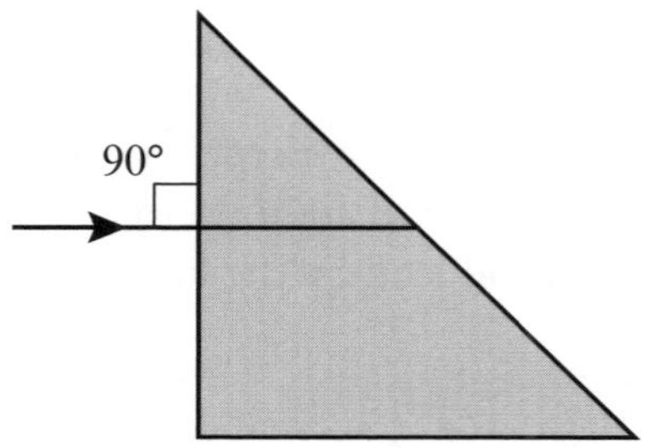

Show that total internal reflection can occur at the sloping face of the prism.

....

....

.... **(2 marks)**

2 Explain how optical fibres make use of total internal reflection in order to transmit light. A labelled diagram should be used to aid your explanation.

....

....

....

.... **(4 marks)**

Exam skills 5
Waves and their properties

1   The figure below shows a ray of light entering a glass block with refractive index 1.50.

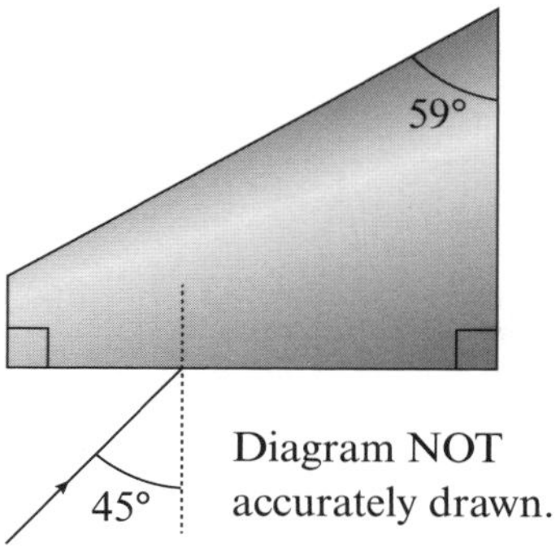

59°

45°

Diagram NOT accurately drawn.

(a)   Determine the angle of refraction as the ray enters the block.

....

....

.... **(2 marks)**

(b)   Explain the term **critical angle** and how it relates to the phenomenon of **total internal reflection**. You should use a diagram to help with your answer.

....

....

....

.... **(3 marks)**

(c)   Determine the critical angle for the glass.

....

....

.... **(2 marks)**

(d)   Determine what happens to the ray when it strikes the sloping face of the block.

....

....

.... **(2 marks)**

(e)   On the diagram at the top of the page, sketch the path that the ray takes until it leaves the block.   **(2 marks)**

Lenses and ray diagrams

1 This question involves drawing accurately scaled ray diagrams in order to determine the position and size of images.

>**Guided**   (a) An object is placed 40.0 cm in front of a converging lens of focal length 15.0 cm:

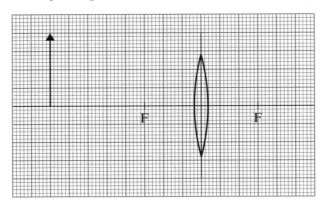

> The image is found using a minimum of two rays: one through the centre of the lens that is unaltered; one parallel to the principal axis that passes through F after passing through the lens.

    (i) Complete the ray diagram to show the image formed.   **(3 marks)**

    (ii) Determine the image distance. .... **(1 mark)**

    (iii) Determine the linear magnification. .... **(1 mark)**

    (iv) Describe the properties of the image.

....

.... **(2 marks)**

   (b) An object is placed 40 cm in front of a diverging lens of focal length 15 cm:

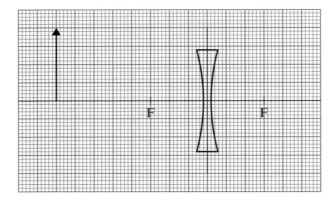

> For a diverging lens, the second ray is drawn as if it came from the principal focus on the **same** side of the lens as the object.

    (i) Complete the ray diagram to show the image formed.   **(3 marks)**

    (ii) Describe the properties of the image.

....

.... **(2 marks)**

Lens formulae

1 Two thin converging lenses A and B have focal lengths of 40.0 cm and 50.0 cm, respectively (see figure below).

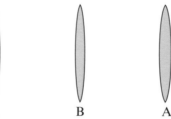

> **Guided**

(a) Determine the power of lens A.

The focal length $f = 40.0\,\text{cm} = 0.400\,\text{m}$. $P = \dfrac{1}{f} =$ ....

.... **(2 marks)**

(b) Determine the combined power of A and B, as in the diagram above.

....

.... **(1 mark)**

(c) Determine the focal length of the combination of A and B.

....

.... **(1 mark)**

2 An object is placed 25.0 cm to the left of a converging lens of focal length 10.0 cm.

(a) Determine the distance of the image from the lens.

....

....

.... **(2 marks)**

(b) The object is moved toward the lens until the object distance is 5.0 cm. Determine the new distance of the image from the lens.

....

....

.... **(2 marks)**

(c) Determine the linear magnification in (b).

....

.... **(1 mark)**

(d) Explain why this image must be virtual.

....

....

> The negative value for the image distance tells you that it is on the same side of the lens as the object.

....

.... **(2 marks)**

Plane polarisation

1 Which of the following waves cannot be polarised?

☐ **A** ultraviolet   ☐ **B** X rays   ☐ **C** ultrasound   ☐ **D** radio   **(1 mark)**

2 The figure shows possible arrangements of two polarising filters.

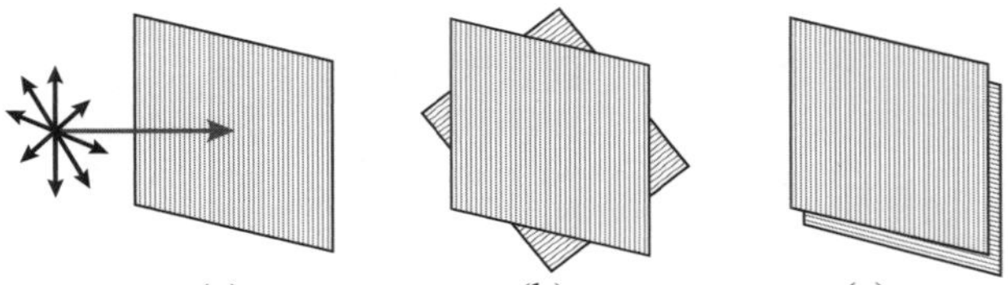

(a)   (b)   (c)

Guided

(a) Explain why a single polarising filter will transmit 50% of incident unpolarised light.

The direction of polarisation of light is actually determined by a vector quantity.

Unpolarised light oscillates in every direction but can be resolved into two

perpendicular components. ....

....

....

.... **(2 marks)**

(b) A second polarising filter is placed on top of the first filter and at 45° to it. Explain why some light will pass through the second filter.

> Think about vectors. There is a component of the polarised light parallel to the transmission direction of the filter.

....

....

.... **(2 marks)**

(c) The second filter is rotated until it is at 90° to the first filter. Explain why no light will emerge from the second filter.

....

....

.... **(2 marks)**

3 Unpolarised light can be partially polarised when it is reflected off a horizontal surface. Explain how sunglasses with polarising lenses can reduce reflections from such surfaces. A diagram can be used to help with your explanation.

....

....

.... **(3 marks)**

Diffraction and Huygens' construction

1   This question concerns Huygens' principle and the use of Huygens' construction in explaining the phenomenon of diffraction.

(a)   State Huygens' principle.

....

.... **(2 marks)**

(b)   Use Huygens' construction to explain diffraction by a narrow slit. Illustrate your answer with a diagram.

> Your diagram should show both the circular 'wavelets' and their sum to give the actual waves.

....

....

.... **(3 marks)**

2   When plane waves, for example light waves, encounter an obstacle, diffraction can also occur, as shown.

Guided

(a)   Using the figure, and adding to it where necessary, explain how bright light can be observed at P.

Light is diffracted by the edges of the

obstacle such that waves are able to

arrive at P. ....

....

.... **(3 marks)**

(b)   Explain why we normally observe shadows when an object is placed between a light source and a screen.

....

....

.... **(2 marks)**

Using a diffraction grating to measure the wavelength of light

1   When light from a laser shines through a diffraction grating, a number of bright fringes can be observed on a screen some distance away.
In an experiment to find the wavelength of the light from a laser, the fringe position x was measured from the zeroth-order fringe to the first- and second-order fringes on each side of the zeroth-order fringe. The distance l was 2.00 m.

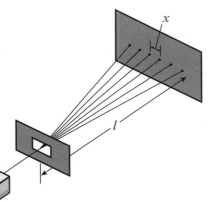

x / mm	1st order	2nd order
left	319	668
right	321	671

⟩Guided⟩   (a)   A grating with 250 lines mm^{-1} is used. Determine the grating spacing, d.

250 lines mm^{-1} means there are 2.50×10^5 lines m^{-1} ....

.... **(1 mark)**

(b)   Determine the mean angle θ subtended by the first-order fringes.

....

.... **(2 marks)**

(c)   Use $n\lambda = d\sin\theta$ to determine the wavelength of the laser used.

....

....

....

.... **(2 marks)**

⟩Guided⟩   (d)   Use the data for the second-order fringes in order to determine the mean value of the wavelength of the laser obtained from the entire experiment.

For $n = 2$, $\theta =$ ....

....

$\lambda = \dfrac{d\sin\theta}{n} =$ ....

.... **(3 marks)**

Practical skills   (e)   Suggest **two** ways in which the wavelength could have been determined more accurately using this apparatus.

....

....

....

.... **(2 marks)**

Electron diffraction

1 Electrons are accelerated by a potential difference of 3.00 kV in an electron gun before striking a graphite target and subsequently forming a pattern of diffraction rings on a phosphor screen (see figure below).

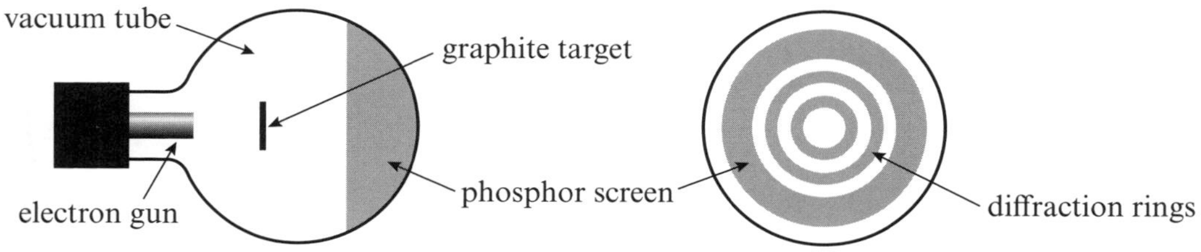

(a) Explain how this experiment provides evidence for the wave nature of electrons.

....

....

.... **(2 marks)**

⟩**Guided**⟩ (b) The target is crystalline graphite. Explain why this results in diffraction rings.

Diffraction only occurs to an appreciable extent when the wavelength of the wave being diffracted is comparable to the spacing in the object doing the diffracting.

....

.... **(2 marks)**

(c) Electrons have mass m and momentum p. Their kinetic energy is equal to:

☐ A $\dfrac{p}{2m}$ ☐ B $\sqrt{\dfrac{p}{2m}}$ ☐ C $\dfrac{p^2}{m}$ ☐ D $\dfrac{p^2}{2m}$ **(1 mark)**

⟩**Guided**⟩ (d) Show that electrons leave the electron gun at about $3 \times 10^7\,\text{m s}^{-1}$.

The electrons of charge e are accelerated by a p.d. V and gain kinetic energy equal

to eV. This means that $eV = \frac{1}{2}mv^2$ and $v = \sqrt{\dfrac{2eV}{m}}$

....

.... **(2 marks)**

(e) Determine the de Broglie wavelength of the electrons.

....

....

> The de Broglie wavelength of a particle is given by $\lambda = \dfrac{h}{p}$ where p is the momentum, mv, and h is the Planck constant.

(1 mark)

(f) The accelerating voltage is increased. The rings will now be:

☐ A of smaller radius and brighter

☐ B of smaller radius and less bright

☐ C of larger radius and brighter

☐ D of larger radius and less bright. **(1 mark)**

Waves and particles

1 This question concerns ideas about the wave or particle nature of electromagnetic waves.

(a) Scientists use different models that try to describe the nature of light. Describe how these ideas have developed over time and how they have contributed to our current understanding.

> Include evidence for and against models of light as particles or as waves.

....

....

....

....

....

.... **(3 marks)**

(b) Name an experiment that provides evidence for the wave nature of light and explain how it does so.

....

....

....

....

....

....

.... **(3 marks)**

(c) Name an experiment that provides evidence for the particle nature of light and explain how it does so.

....

....

....

....

....

....

.... **(3 marks)**

Guided 2 Electrons, often considered as particles, can also exhibit wave properties. Determine the de Broglie wavelength of electrons travelling at $3.0 \times 10^7 \, \text{m s}^{-1}$.

The de Broglie wavelength of a particle is given by $\lambda = \dfrac{h}{p} = $ ....

.... **(2 marks)**

The photoelectric effect

1   An experiment to demonstrate the photoelectric effect uses a gold-leaf electroscope like the one shown. The gold-leaf electroscope has a polished zinc plate placed on its cap, which is then negatively charged. When the zinc plate is illuminated with ultraviolet light, the gold-leaf electroscope is discharged. This does not happen when visible light of any intensity is used.

zinc plate

gold-leaf electroscope

(a)   Explain why the experiment only works if the gold-leaf electroscope is initially charged negatively.

....

....

.... **(2 marks)**

(b)   Explain why the gold-leaf electroscope is **not** discharged when illuminated with visible light.

....

.... **(2 marks)**

(c)   Explain why the gold-leaf electroscope is discharged when illuminated with ultraviolet light.

....

....

.... **(2 marks)**

(d)   Explain why, when a metal other than zinc is used on the top plate of the electroscope, the threshold frequency changes.

....

....

.... **(2 marks)**

2   The work function of sodium is 2.36 eV.

(a)   Define the work function, ϕ, of a metal.

....

.... **(1 mark)**

(b)   Determine the threshold frequency of sodium.

....

....

> You need to convert the energy in electronvolts into joules with the conversion factor from the data sheet.

(2 marks)

(c)   Determine the maximum kinetic energy of a photoelectron emitted from the surface of a piece of sodium that is illuminated by light of wavelength 465 nm.

....

.... **(2 marks)**

Line spectra and the eV

1 Describe the process that results in the emission of light in the form of an atomic line spectrum from a substance such as a hot vapour.

....

....

....

....

.... **(4 marks)**

2 The figure (not to scale) shows some of the energy levels predicted using Niels Bohr's model of the hydrogen atom.

The energy levels, in electronvolts, are calculated using the formula

$E = \dfrac{-13.6}{n^2}$, where $n = 1, 2, 3$, etc.

(a) Determine the energy levels in electronvolts for $n = 4$ and $n = 5$ and write your answers in the boxes provided.   **(1 mark)**

> **Guided**

(b) Determine the lowest energy state of hydrogen in joules.

$-13.60 \times 1.60 \times 10^{-19} =$ ....

.... **(1 mark)**

(c) Determine the energy of a photon produced when an electron moves from the energy level corresponding to that corresponding to $n = 3$ to $n = 2$.

....

.... **(2 marks)**

> **Guided**

(d) Determine the wavelength of a photon produced by the process in (c) above.

$\Delta E = hf = \dfrac{hc}{\lambda}$, which gives $\lambda = \dfrac{hc}{\Delta E}$, where ΔE is the energy calculated in (c)

....

.... **(2 marks)**

(e) Show that if a photon with a wavelength of 103 nm is **absorbed**, an electron can be raised in energy from the $n = 1$ energy level to the $n = 3$ energy level.

....

....

....

.... **(3 marks)**

The figure shows energy levels:
0 eV — $n = 5$, $n = 4$
-1.51 eV — $n = 3$
-3.40 eV — $n = 2$
-13.60 eV — $n = 1$

Exam skills 6
The quantum nature of light

1　This questions concerns the photoelectric effect and Einstein's equation.

(a)　Explain the meaning of the term **threshold frequency** in the context of the photoelectric effect.

....

....

....　**(2 marks)**

The sketch graph below shows how the kinetic energy of photoelectrons varies with the frequency of incident light on a metal surface.

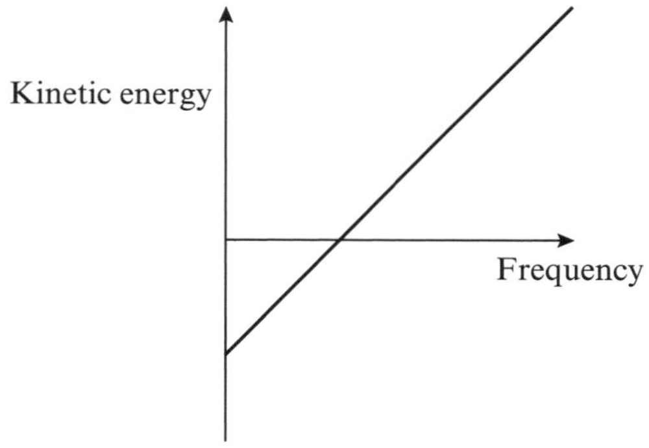

(b)　Label the threshold frequency, f_0.　**(1 mark)**

(c)　Explain how the threshold frequency is related to the work function of the metal used.

....

....

....　**(2 marks)**

(d)　Add a second line to the graph that represents the behaviour of a metal with a higher work function.　**(2 marks)**

Einstein's equation states that the energy of incident photons is equal to the work function added to the maximum kinetic energy of emitted photoelectrons, or $hf = \phi + KE_{max}$

(e)　Describe the nature of a photon.

....

....

....　**(2 marks)**

(f)　Explain why the final term in the equation is the **maximum** kinetic energy of emitted photoelectrons.

....

....

....　**(2 marks)**

Impulse and change of momentum

1 A 'superball' of mass 0.20 kg falls vertically onto a smooth horizontal surface and bounces vertically upwards. The collision is elastic. The figure shows how the upward force on the ball varies during its time of contact with the ground.

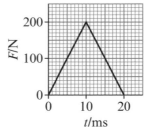

> If a collision is 'elastic' then kinetic energy is conserved as well as linear momentum.

(a) State the name of the physical quantity represented by the area under the graph and give its S.I. unit.

.... **(2 marks)**

›Guided›

🖩 Maths skills

(b) Calculate the change of momentum of the ball during the impact.

The area under the graph ....

....

....

(3 marks)

(c) Calculate the velocity of the ball just before it hits the ground.

....

....

....

....

> Do not forget that the ball bounces, so the total change of momentum includes downward velocity dropping to zero and then upward acceleration.

(3 marks)

2 The advertising material for a company supplying surfaces for children's playgrounds claims that: 'Our surface is designed to cushion the impact of a child's accidental fall and reduce the severity of any injury.' The material supplied compresses under impact. Explain how such a material can 'reduce the severity of any injury'.

....

....

....

....

> Think of the physics and refer to physical principles – Newton's second law is important here. Try to identify four distinct points.

(4 marks)

Conservation of momentum in two dimensions

1 (a) State the law of conservation of momentum and explain the conditions under which it applies.

....

....

> 'Explain' means you must give more detail than just stating the condition.

.... **(3 marks)**

(b) A spacecraft of mass 10 000 kg is travelling at a constant velocity of $3.2 \times 10^4 \, \text{m s}^{-1}$. It makes an adjustment to its course by firing a rocket for 30 s. The average thrust from the rocket is $2.0 \times 10^6 \, \text{N}$ in a direction at 90° to the original direction of travel of the rocket.

(i) Calculate the change of momentum of the spacecraft during the 30 s the rocket is fired.

....

.... **(2 marks)**

(ii) Calculate the angle between the spacecraft's initial direction and its new path as a result of firing the rocket. You might find it helpful to draw a diagram.

> If the question suggests drawing a diagram it is usually a good idea to do so! In this case a vector diagram of the momenta would be useful. It is a right-angled triangle.

....

....

.... **(3 marks)**

(iii) What can you say about the momentum of the gases ejected by the rocket?

....

.... **(2 marks)**

Guided 2 In radioactive beta decay, a high-speed electron and a neutrino are ejected from the nucleus of an unstable atom. A neutrino is also emitted. Both particles have momentum. The diagram shows the directions in which the electron and the neutrino are ejected from an unstable atom that is initially at rest. Add a third arrow to show a direction in which the nucleus could recoil, and explain your answer.

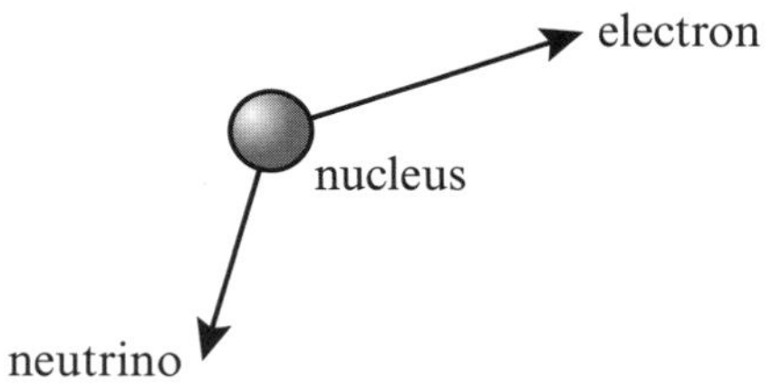

Momentum is always conserved, so ....

....

.... **(3 marks)**

Elastic and inelastic collisions

1   A man of mass 70 kg running at 8.0 m s^{-1} jumps into a stationary boat of mass 230 kg and the boat begins to move forwards. All motions are horizontal and drag forces can be ignored.

(a)   Discuss without calculation how the law of conservation of energy and the law of conservation of momentum apply to this event.

> There are four marks available, so you should aim to make two points about each conservation law.

....

....

....

.... **(4 marks)**

Guided

(b)   Show that the velocity of the boat just after the man lands inside it is about 1.9 m s^{-1}.

By conservation of momentum, velocity of man and boat v_2 = ...

....

....

.... **(3 marks)**

(c)   (i)   Calculate the kinetic energy of the man before he lands in the boat and of the man and boat afterwards.

....

....

.... **(2 marks)**

(ii)   State and explain whether this is an elastic or inelastic collision.

....

....

.... **(2 marks)**

(iii)   The boat drifts for a while and eventually stops.
What has happened to its kinetic energy and linear momentum?

....

....

.... **(2 marks)**

Investigating momentum change

1   The figure below shows the apparatus used to investigate the relationship between impulse and momentum change.

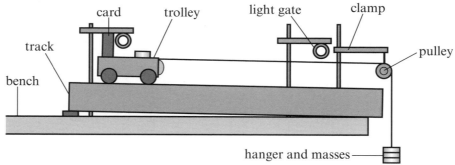

(a)   Before taking any measurements the track is tilted slightly as shown in the diagram. Suggest and explain a reason for this.

....

....

.... **(3 marks)**

(b)   Explain how the resultant force on the trolley is varied.

....

.... **(2 marks)**

(c)   In a particular experiment, the hanger and all of the masses have the same mass. In order to vary the resultant force, the experimenter starts with several masses on the hanger and then repeats her measurements as she moves one mass at a time from the hanger until just the hanger remains. Each mass that she removes from the hanger is placed on top of the trolley. Explain why she does this.

....

.... **(2 marks)**

(d)   How will she use the results of this experiment to calculate the impulse given to the trolley?

....

.... **(2 marks)**

(e)   How will she use the results of this experiment to calculate the momentum change?

....

.... **(2 marks)**

(f)   Having calculated and tabulated values for the impulse and change of momentum she then plots a graph of change of momentum against impulse. What gradient should she expect from this graph? Explain your answer.

....

....

.... **(3 marks)**

Exam skills 7 Impulse and change of momentum

1   The figure shows how the driving force and total drag force on a car varies with time as it pulls away from a standstill at a road junction. The mass of the car and driver is 1400 kg.

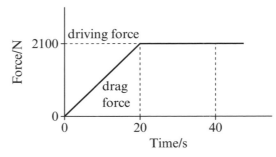

   (a)   Sketch, on the axes shown, a graph to show how the resultant force on the car varies with time during this motion.

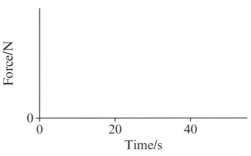

       (3 marks)

   (b)   Calculate the impulse given to the car during the motion.

....

.... **(2 marks)**

   (c)   Calculate the velocity of the car after 40 s.

....

.... **(2 marks)**

   (d)   Sketch, on the axes shown, a graph to show how the velocity of the car changes with time during the motion.

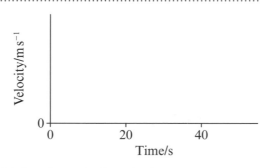

       (3 marks)

   (e)   Show that the total work done by the driving force during the first 40 s of the motion is about 950 kJ.

....

....

.... **(3 marks)**

   (f)   Show that the work done against frictional forces during the first 40 s of the motion is about 790 kJ.

....

....

.... **(3 marks)**

Describing rotational motion

1   A washing machine drum is rotating at constant angular velocity ω.
Which of the following statements about a point on the drum is correct?

     ☐ **A** The time for it to complete one rotation is $2\pi\omega$.

     ☐ **B** Its velocity is constant.

     ☐ **C** It is accelerating.

     ☐ **D** There is a resultant outward force acting on it.   **(1 mark)**

Maths skills

2   The table contains data about the Earth and Mars. Assume that the orbits of Earth
and Mars are circular. There are 3.16×10^7 s in one year.

Planet	Mean distance from Sun, r/million km	Orbit time, T/years
Earth	150	1.0
Mars	228	

   (a) Calculate the angular velocity of the Earth as it orbits
the Sun. Give your answer in radians per second.

> Make sure you convert km to m and years to seconds.

....

....

....   **(3 marks)**

   (b) Calculate the Earth's orbital speed in m s^{-1}.

....

....   **(1 mark)**

Guided

   (c) (i) The astronomer Johannes Kepler proposed three laws of planetary motion.
His third law states that the ratio of $\dfrac{r^3}{T^2}$ is constant for all planets in the
Solar System. Use this law to calculate the orbital period of Mars, which is
missing from the table above.

From Kepler's third law $\dfrac{r_E^3}{T_E^2} = \dfrac{r_M^3}{T_M^2}$ ....

....

....   **(3 marks)**

       (ii) Calculate the angular displacement of Mars in its orbit during one Earth year.
Give your answer in both radians and degrees.

....

....

....   **(3 marks)**

Uniform circular motion

1   A child on a playground roundabout is moving in uniform circular motion. He is sitting 2.5 m from the centre of the roundabout and completes one revolution every 6.0 s.

(a) Use a vector diagram to explain why the child must have an acceleration toward the centre of the circle.

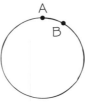

During a short time the child will move from A to B and the velocity vector will

change direction. Acceleration is defined as ....

....

....

.... **(4 marks)**

(b) Calculate the centripetal acceleration of the child.

....

.... **(2 marks)**

(c) (i) Explain why there must be a resultant force acting on the child and state its direction.

Think about $F = ma$ and relate it to part (a).

....

....

.... **(2 marks)**

(ii) Suggest what could provide this resultant force.

....

.... **(1 mark)**

(d) The child stays on the roundabout for 25 s. Calculate the angular displacement in radians moved by the child in this time.

....

....

.... **(3 marks)**

Centripetal force and acceleration

1 A racing car of mass 1200 kg is cornering at constant speed of $18\,\text{m s}^{-1}$ and follows a path that is part of a horizontal circle of radius 45 m.

(a) Calculate the magnitude of the resultant force acting on the car and state its direction.

....

.... **(2 marks)**

(b) Explain how this resultant force is produced.

....

.... **(2 marks)**

(c) On the next lap, the driver of the car attempts to take the corner at a higher speed.
Explain why the car might skid off the track.

> There is no outward force acting on the car, but the inward force is limited.

....

....

.... **(2 marks)**

2 A simple pendulum consists of a light inextensible string of length 0.50 m and a small bob of mass 0.060 kg. The bob is displaced sideways so that it rises through a vertical height of 0.050 m and is then released.

Guided

(a) Explain why the tension in the string at the moment the bob passes through its lowest position is greater than the weight of the bob.

When the bob moves through its lowest position, it is moving in circular motion,

so there must be a resultant force toward ....

....

.... **(3 marks)**

(b) Calculate the tension in the string at the moment the bob passes through its lowest position.

> Sketch a free-body diagram of the bob as it passes the lowest position and use this to find an expression for the tension force.

....

....

....

.... **(4 marks)**

Electric field strength

1 (a) State what is meant by an **electric field** and define **electric field strength**.

....

....

.... **(2 marks)**

(b) The air breaks down and becomes conducting when the electric field exceeds
about $3 \times 10^6\,V\,m^{-1}$. In an experiment to demonstrate electrostatic effects,
a teacher gradually increases the potential difference between two metal
conductors until a spark jumps between them. The separation of the conductors
is 2.5 mm. What is the approximate value of the minimum potential difference
needed to produce the spark?

....

.... **(2 marks)**

(c) A capacitor consists of two parallel metal plates separated by a distance of
4.0 mm. A fixed potential difference of 500 V is applied to the plates and they
are gradually moved farther apart until their separation is 8.0 mm.

(i) Calculate the electric field strength between the plates when they are
4.0 mm apart.

.... **(1 mark)**

(ii) Calculate the electric field strength between the plates when they are
8.0 mm apart.

.... **(1 mark)**

(iii) Sketch a graph to show how the electric
field strength between the plates varies
with their separation while the potential
difference between them remains at 500 V,
and label it with the values calculated above.

(3 marks)

2 Draw a diagram to show the electric field lines around the two separate identical
positive charges shown below.

(3 marks)

Electric field and electric potential

1   An electron emitted from an electron gun is accelerated through a potential difference
    of 420 V in a vacuum.
    The charge on an electron is -1.6×10^{-19} C. The mass of an electron is 9.1×10^{-31} kg.

(a) Calculate the work done on the electron.

....

....

....   **(2 marks)**

(b) Calculate the final velocity of the electron.

....   In a vacuum, all of the
                                                                                        electrical potential
....   energy is transferred
                                                                                        to kinetic energy.

....

....   **(3 marks)**

2   The potential difference between the ground and the base of a thunder cloud is about
    10^9 V just before a lightning strike occurs. The base of the cloud is 300 m above the
    ground and the lightning strike transfers 16 C of charge to the ground.

(a) Calculate the mean value of the electric field strength beneath the cloud.

....

....   **(2 marks)**

(b) Calculate the energy released by the lightning strike and explain what happens
     to this energy.

....

....

....   **(3 marks)**

3   Complete the diagram below by adding a direction to the field lines and drawing three
    equipotentials. Label the equipotential with the highest potential.

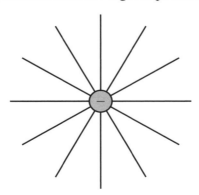

(4 marks)

Forces between charges

1   Two particles, each carrying a charge Q, are separated by a distance d. The electrostatic force between them is F. What is the electrostatic force between two different particles, which each carry a charge $2Q$ and are separated by a distance $2d$?

☐ **A** $\dfrac{F}{4}$ ☐ **B** $\dfrac{F}{2}$ ☐ **C** F ☐ **D** $2F$ **(1 mark)**

2   Many molecules are dipoles, positive at one end and negative at the other. A simple model of a dipole consists of two point charges $+Q$ and $-Q$ separated by a distance $2r$, as shown in the figure.

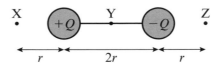

(a)   State the direction of the force on an electron at X, Y (the mid-point of the dipole) and Z (these are all points on the line through Q and $-Q$).

....

....

| Think about the resultant force. |

.... **(3 marks)**

> **Guided**

(b)   Calculate the force on an electron at Y (halfway between the two charges). Assume that $Q = 8.0 \times 10^{-20}$ C. The charge on an electron is 1.6×10^{-19} C; $r = 2.5 \times 10^{-10}$ m; $\varepsilon_0 = 8.85 \times 10^{-12}$ F m^{-1}.

Taking forces to the right as positive, the resultant force on the electron is (by Coulomb's law)

$$F = \frac{Qe}{4\pi\varepsilon_0 r^2} - \left(-\frac{Qe}{4\pi\varepsilon_0 r^2}\right) = \text{..}$$

....

....

.... **(4 marks)**

(c)   Calculate the force on an electron at Z.

| The electron will be affected by both $+Q$ and $-Q$. |

....

....

....

.... **(4 marks)**

(d)   Some solid materials contain dipole molecules. Suggest how these molecules might behave when an external electric field is applied to the solid.

....

....

.... **(2 marks)**

Field and potential for a point charge

1 The spherical dome of an electrostatic generator has a radius of 0.20 m and is raised to a potential of 400 kV with respect to Earth. ($\varepsilon_0 = 8.85 \times 10^{-12}$ F m^{-1}.)

>**Guided**

(a) Calculate the charge stored on the dome.

The surface of the sphere is like an equipotential surface around a point charge, so

$V =$ ....

....

| For this question you can use the equation for a point charge. |

....

.... **(3 marks)**

(b) What is the potential 0.20 m from the surface of the dome?

| 🖩 Maths skills   Potential is inversely proportional to distance. |

....

.... **(2 marks)**

(c) Calculate the electric potential 5.0 cm above the surface of the dome.

....

....

.... **(2 marks)**

(d) (i) State the relationship between electric field strength and electric potential.

....

.... **(1 mark)**

(ii) Calculate the electric field strength immediately above the surface of the dome.

....

....

.... **(2 marks)**

(iii) What is the electric field strength 0.20 m from the surface of the dome?

| 🖩 Maths skills   Electric field strength obeys an inverse-square law. |

....

....

.... **(2 marks)**

(e) The dome itself is a hollow conductor, and the electric field strength everywhere inside the dome is zero. State and explain the value of potential at the centre of the dome.

| Think about your answer to (d)(i). |

....

....

.... **(2 marks)**

Capacitance

Guided 1 Which of the following combinations of units is **not** equivalent to the farad?

    ☐ **A** $A\,s\,J^{-1}$

    ☐ **B** $C\,V^{-1}$ ◄

    ☐ **C** $C^2\,J^{-1}$

    ☐ **D** $s\,\Omega^{-1}$

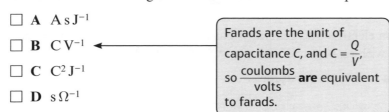

Farads are the unit of capacitance C, and $C = \dfrac{Q}{V}$, so $\dfrac{\text{coulombs}}{\text{volts}}$ **are** equivalent to farads.

(1 mark)

2 A 220 μF capacitor is charged from a 12 V supply through a 100 Ω resistor using the circuit shown below.

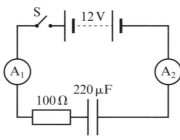

(a) How much charge is stored on the positive capacitor plate when it is fully charged?

....

.... **(2 marks)**

(b) What current flows through each ammeter at the moment just after S is closed?

....

.... **(2 marks)**

(c) Explain why the current will decrease as the capacitor charges.

....

....

.... **(2 marks)**

Guided (d) Calculate the potential difference across the 100 Ω resistor when the capacitor is 80% charged and the charging current at that time, and explain your reasoning.

> Approach this logically: what voltage is across the capacitor when it is 80% charged? What voltage does this leave for the resistor? Then use $I = \dfrac{V}{R}$ for the resistor.

80% charge $Q = 0.8 \times CV$ (part a) = ...

so p.d. across capacitor $V = \dfrac{Q}{C}$ = ...

Therefore, p.d. across resistor is ....

Current in circuit when capacitor is 80% charged $I = \dfrac{V}{R}$ = ...

(4 marks)

Energy stored by a capacitor

1 The graph shows how the charge stored on a capacitor increases as the potential difference across it is increased.

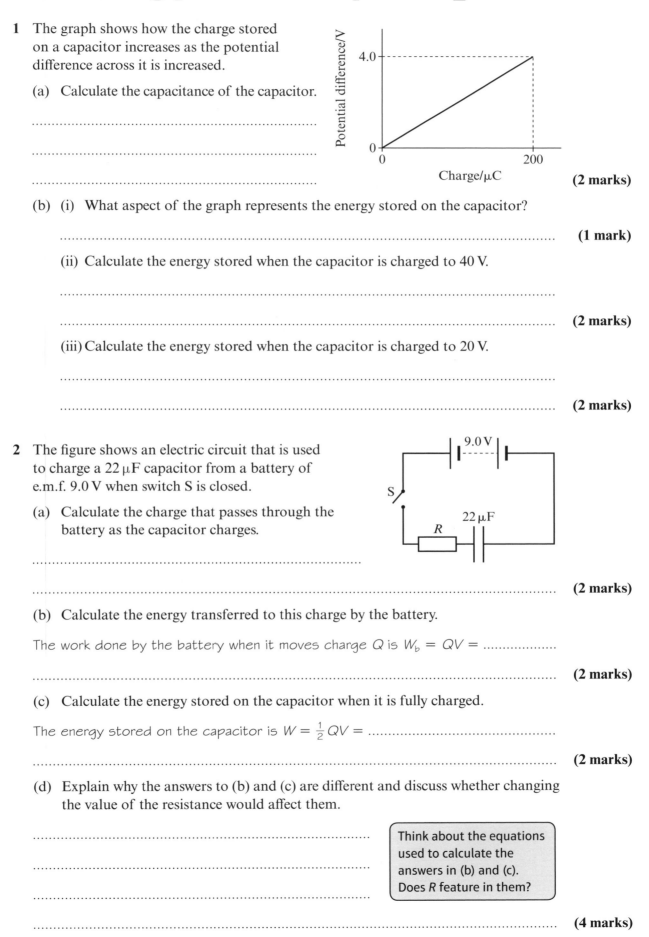

(a) Calculate the capacitance of the capacitor.

....

....

....

(2 marks)

(b) (i) What aspect of the graph represents the energy stored on the capacitor?

....

(1 mark)

(ii) Calculate the energy stored when the capacitor is charged to 40 V.

....

....

(2 marks)

(iii) Calculate the energy stored when the capacitor is charged to 20 V.

....

....

(2 marks)

2 The figure shows an electric circuit that is used to charge a 22 μF capacitor from a battery of e.m.f. 9.0 V when switch S is closed.

> Maths
> skills

(a) Calculate the charge that passes through the battery as the capacitor charges.

....

....

(2 marks)

> Guided

(b) Calculate the energy transferred to this charge by the battery.

The work done by the battery when it moves charge Q is $W_b = QV = $ ....

....

(2 marks)

> Guided

(c) Calculate the energy stored on the capacitor when it is fully charged.

The energy stored on the capacitor is $W = \frac{1}{2}QV = $ ....

....

(2 marks)

(d) Explain why the answers to (b) and (c) are different and discuss whether changing the value of the resistance would affect them.

....

....

....

> Think about the equations
> used to calculate the
> answers in (b) and (c).
> Does *R* feature in them?

....

(4 marks)

Had a go ☐ Nearly there ☐ Nailed it! ☐

Charging and discharging capacitors

1 Identify the graph below, which applies to both capacitor charge and capacitor discharge, by ticking the correct letter.

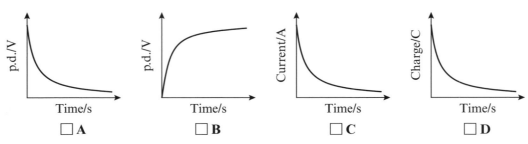

☐ **A** ☐ **B** ☐ **C** ☐ **D** **(1 mark)**

2 (a) A 400 μF capacitor is fully charged from a 9.0 V supply. Calculate the charge it stores.

.... **(1 mark)**

(b) Sketch a circuit diagram for a circuit that could be used to observe the charging of a capacitor and record observations from which to plot graphs of p.d. and current against time.

> You need separate sensors to log the data for p.d. and current.

(3 marks)

Guided

(c) A capacitor is connected in series with a fixed resistance and charged from a constant supply. Describe and explain how the charge on the capacitor plates, the current in the circuit and the potential difference across the capacitor changes with time.

As the electrons carrying charge around the circuit accumulate on the negative plate and drain away from the positive plate of the capacitor, ...

....

....

....

....

.... **(6 marks)**

The time constant

1 (a) Explain the significance of the time constant $\tau = RC$ in a capacitor charging circuit.

....

....

.... **(3 marks)**

(b) Show that the unit of resistance × the unit of capacitance is the unit of time.

....

....

....

> **Maths skills** Units in all physical equations must balance. Use equations you know to reduce complex units such as the ohm to simpler combinations (e.g. $1\,\Omega = 1\,V\,A^{-1}$).

(2 marks)

2 A capacitor is charged using the circuit shown.

(a) Calculate the time constant for this circuit.

....

....

....

[Circuit diagram: 24 V supply, switch S, 220 Ω resistor, 100 μF capacitor]

(2 marks)

> **Guided**

(b) S is closed and the capacitor begins to charge. Approximately how long will it take for the capacitor to reach 99% of its full charge?

$(0.37)^5 = 0.007$, that is, in five time constants a discharged capacitor will be 99.3% charged, so

....

....

> A useful rule of thumb is that over one time constant the charge on a charging or discharging capacitor changes by 37%.

(2 marks)

(c) The time constant and, therefore, the time taken to charge the capacitor increase if the capacitance or resistance is increased. Explain why this is.

....

....

....

.... **(4 marks)**

3 A charged capacitor of capacitance 50 μF is discharged through a resistor of resistance 20 kΩ. Calculate the time taken for the charge on the capacitor to fall to approximately 1% of its initial value.

....

....

> First convert the units μF and kΩ to F and Ω.

.... **(2 marks)**

Exponential decay of charge

1   The $1000\,\mu\text{F}$ capacitor shown in the figure below is charged for $3.0\,\text{s}$ and then discharged through a $500\,\Omega$ resistor.

(a) Use the axes below to sketch a graph to show how the potential difference across the capacitor changes with time during the charging and discharging cycle.

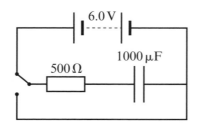

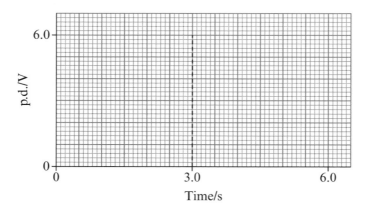

Calculate the time constant and use this as a 'yardstick' for the time scale. How long does the capacitor take to charge or discharge?

(6 marks)

(b) Calculate the charge on the capacitor when it is fully charged.

....

.... **(1 mark)**

(c) Calculate the maximum charging current and state when it occurs.

....

....

.... **(2 marks)**

⟩Guided⟩ (d) How long will it take for the charge on the discharging capacitor to fall from its initial value to half that value?

$Q = Q_0\, e^{-\frac{t}{RC}}$ so when charge is halved,

$\dfrac{Q}{Q_0} = e^{-\frac{t}{RC}} = 0.5$

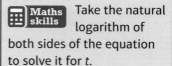

Take the natural logarithm of both sides of the equation to solve it for t.

....

....

.... **(3 marks)**

(e) Add a line to your graph to show how the potential difference across the resistor changes during the charging and discharging cycle.

(2 marks)

Exam skills 8 Capacitors

1 The circuit shown in the figure is a simplified model of the circuit used in a camera flash gun. The lamp emits light when the potential difference across it is greater than 60 V. The sequence of events to make the lamp flash is:

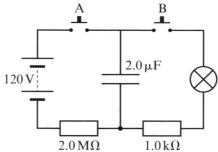

1. Close switch A for at least 12 s.
2. Open switch A.
3. Close switch B.
4. Open switch B.

(a) Explain why it is recommended to close switch A for at least 12 s.
Support your answer with a relevant calculation.

....

....

....

.... **(3 marks)**

(b) Calculate the energy stored on the capacitor when:

   (i) it is fully charged by the 120 V supply.

   ....

   .... **(2 marks)**

   (ii) its voltage has fallen to 60 V.

   ....

   .... **(1 mark)**

(c) Calculate the time taken for the voltage across the capacitor to fall from 120 V to 60 V during its discharge.

....

.... **(2 marks)**

(d) Use your answers to (b) and (c) to estimate the power output of the flash lamp.

....

....

....

.... **(4 marks)**

(e) The owner of the camera notices that when he uses the flash twice in succession it takes less time to charge after the first use. Suggest a reason for this.

....

....

.... **(2 marks)**

Describing magnetic fields

1 Which of the following statements about magnetic fields is correct?

☐ **A** The two ends of a bar magnet carry opposite charges.

☐ **B** Like poles attract one another.

☐ **C** Magnetic field lines go from south to north.

☒ **D** The stronger the magnetic field, the denser the field lines. **(1 mark)**

2 A uniform coil of N turns and area A lies with its plane parallel to a uniform magnetic field of strength B. Which line below correctly gives the flux ϕ through the coil?

☒ **A** $\phi = 0$

☐ **B** $\phi = B$

☐ **C** $\phi = BA$

☐ **D** $\phi = NBA$ **(1 mark)**

3 The Earth's magnetic field strength at a particular position near the surface is $36\,\mu T$ at an angle of $30°$ to the horizontal.

(a) Calculate the magnetic flux through $2.0\,m^2$ of the Earth's surface.

> **Maths skills** You will have to resolve a vector to answer this question.

$$36 \times 10^{-6} \times \cos(30)$$
$$= 3.11 \times 10^{-5}\,T \times 2 = 6.2 \times 10^{-5}\,wb$$ **(3 marks)**

Guided

(b) A large coil of wire has a cross-sectional area of $0.25\,m^2$ and consists of 500 turns of insulated wire. The coil is placed so that its plane is vertical to the ground and perpendicular to the horizontal component of the Earth's magnetic field. Calculate the magnetic flux linkage through the coil.

$B_{horizontal} = 36 \times 10^{-6} \times \cos 30° = 3.11 \times 10^{-5}\,T$

$NB_h A = 3.11 \times 10^{-5} \times 0.25 \times 500 = 3.89 \times 10^{-3}$

$3.89 \times 10^{-3}\,wb$ **(3 marks)**

4 Explain what is meant by a **uniform magnetic field** and discuss when it is possible to treat the Earth's magnetic field as uniform.

- Uniform MF is when the field lines run parallel to eachother from north to south
- It is possible to treat earth MF as uniform when you are standing on the equator
- Because the field lines will be parallel. **(4 marks)**

Forces on moving charges in a magnetic field

1 An electron moving at velocity v is fired into a uniform magnetic field acting into the page as shown in the figure.

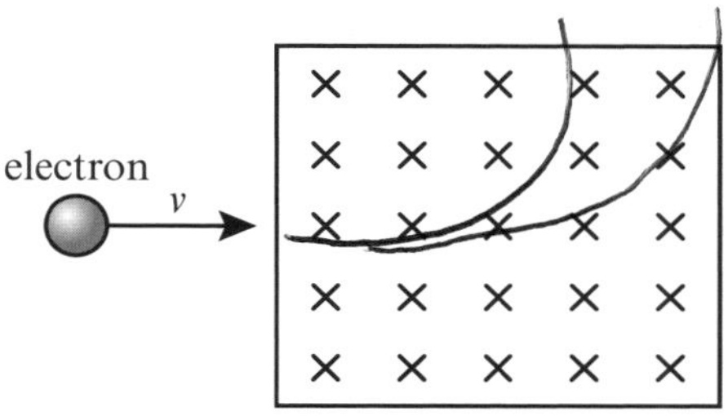

electron

v

(a) Draw a line on the diagram to show the path of the electron inside the field. State the shape of the path.

.... Curve . ... **(2 marks)**

(b) Add a second line to show the path of a second electron that enters the field along the same line but with velocity $2v$. Explain why this path differs from the path of the first electron.

~ Because it spends less time over the MF
~ So it is acted on by by a force for less time .... **(2 marks)**

(c) A third electron enters the field moving parallel to the magnetic field lines (into the page). Describe and explain the path of this electron in the field.

~Moves in straight line.
~ Because electron not experiencing change in magnetic ...... **(2 marks)**
    flux.

⮞Guided⮞ 2 A magnetic field can be used to separate isotopes, for example, to separate fissile uranium-235 from non-fissile uranium-238. The first stage is to accelerate ions of the mixture up to the same speed. These ion beams are then injected into a uniform magnetic field at right angles to their motion. Inside the field, they move in circular paths of different radii. In a particular experiment, the ions have a velocity v and the field has strength B. The masses of the two ions are 235 u and 238 u (where u is the unified atomic mass unit) and they each carry a charge q.

(a) Derive an expression for the separation of the ion beams after they have turned through a semicircle in the magnetic field.

> 1. Write down an expression for the radius of the path for each ion (r_1 and r_2)
> 2. When they have moved through a semicircle in the field they will be two radii from their starting point.
> 3. Their separation is $2(r_1 - r_2)$.

....

....

....

....

.... **(4 marks)**

(b) Suggest ways to increase the separation of the ion beams.

....

.... **(2 marks)**

Electromagnetic induction – relative motion

1   The figure shows a simple A.C. generator consisting of a coil of 120 turns and area 5.0 cm² rotating in a uniform magnetic field of strength 40 mT. The coil rotates at a steady rate, completing one rotation in T seconds.

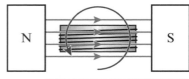

Watch out for a change of sign of flux linkage.

(a)   At $t = 0$, the coil is in the position shown and the flux linkage is increasing. Calculate the flux linkage through the coil at the following times:

(i) $t = 0$

.... 0 ....

(ii) $t = 0.25\,T$

.... 2.4×10^{-3} ....

(iii) $t = 0.50\,T$

.... 0 ....

(iv) $t = 0.75\,T$

.... $\leftarrow 2.4 \times 10^{-3}$ ....   **(4 marks)**

(b)   Hence sketch a graph to show how the flux linkage varies with time during one complete rotation of the coil. Add a suitable scale and unit to the y-axis.

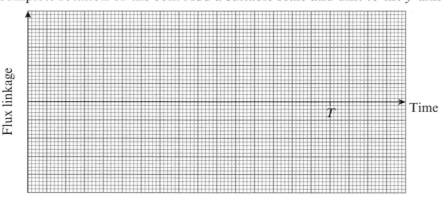

(4 marks)

(c)   (i)   Use the graph from part (a) to sketch a graph of how the induced e.m.f. in the coil varies during one rotation. Do not put a scale on the y-axis.

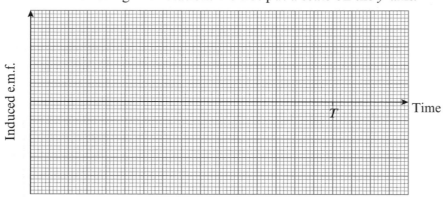

(2 marks)

(ii)   Explain the relationship between the graphs of e.m.f. and flux linkage.

....

....

....

....   **(3 marks)**

Maths skills   Think about Faraday's law – this is all to do with rates of change or gradients.

Changing flux linkage

1 The figure shows an experimental set-up similar to one used by Michael Faraday when he investigated electromagnetic induction.

When the switch is closed, the compass needle deflects to the left and then returns to its original position.

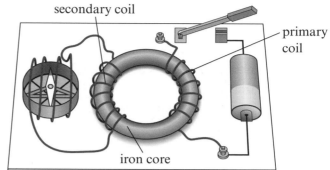

secondary coil

primary coil

iron core

(a) Explain, in as much detail as you can, why the needle deflects and returns.

When the switch is closed, the current in the primary coil increases from zero.

This causes an increase in the magnetic flux in the iron core. This changing flux

links the secondary coil and induces ....

....

....

....

> You could answer a long question like this by giving a series of bullet points, but there must be a clear logical link joining each point.

....

....

....

....

....

....

....   **(8 marks)**

(b) Explain the purpose of the iron core.

....

....   **(2 marks)**

(c) The switch is opened. State and explain what, if anything, happens to the compass needle.

....

....

> The magnetic field inside the core collapses – think about Faraday's law.

....

....

....   **(4 marks)**

(d) Explain why transformers must be used with an A.C. supply.

....

....

....   **(3 marks)**

Faraday's and Lenz's laws

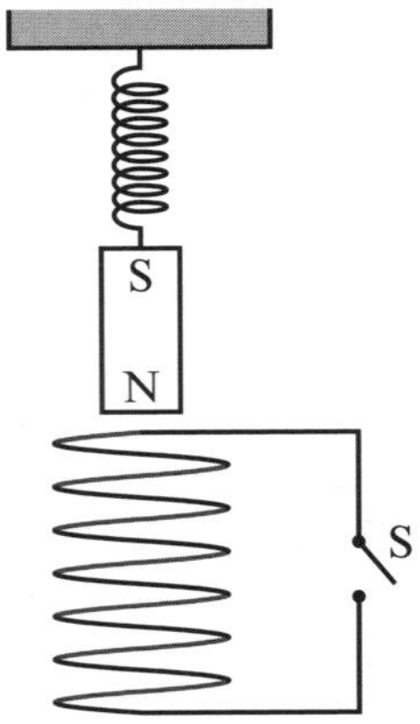

1   A bar magnet is suspended from a spring above a coil as shown in the figure.
With S open, the magnet is displaced and oscillates freely up and down. While it is oscillating S is closed and the oscillations rapidly die away.

(a) Use Faraday's law to explain why there will be an induced e.m.f. but no induced current in the coil before the switch is closed.

....

....

....

....

....

> Use bullet points to make three separate, linked points.

(3 marks)

(b) Use Lenz's law to explain why the oscillations die away when the switch is closed.

....

....

....   **(3 marks)**

(c) Lenz's law is a consequence of the law of conservation of energy.
Explain how this example illustrates the law of conservation of energy.

....

....

....   **(2 marks)**

2   A flat coil of 50 turns and area $25\,cm^2$ lies so that its plane is parallel to a uniform magnetic field of strength $20\,mT$. The coil is then rotated through 90° in a time of $0.20\,s$ so that it finishes with its plane perpendicular to the field.

> Guided

(a) Calculate the change in flux linkage through the coil as it is rotated through 90°.

change in flux linkage = $NB\Delta A$ = ...   **(2 marks)**

(b) Calculate the average induced e.m.f. in the coil during the rotation.

....

....   **(3 marks)**

Alternating currents

1 The figure shows a graph of voltage against
   time for the output of an A.C. power supply
   used in a school laboratory.

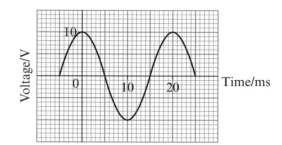

(a) Explain the difference between an A.C.
     supply and a D.C. supply.

     ....

     .... **(2 marks)**

(b) State the following values for the supply:

     Peak value of A.C. voltage: ....

     Root mean square (r.m.s.) value of A.C. voltage: ....

     Time period of A.C. supply: ....

     Frequency of A.C. supply: .... **(4 marks)**

Guided

(c) The supply above is connected to a $100\,\Omega$ resistor.

     (i) Calculate the r.m.s. value of the current through the resistor.

          $\dfrac{V_{rms}}{R} = I_{rms}$ ....

          .... **(1 mark)**

     (ii) Calculate the average power transferred by the resistor when it is connected
          to this power supply.

          Average power $= V_{rms}I_{rms} =$ ....

          .... **(2 marks)**

(d) (i) A 9.0 V battery is used to light a filament lamp. The light glows
          brightly. Describe and explain how the lamp will glow if it is
          connected to the A.C. supply above.

Power in a D.C. circuit is $VI = \dfrac{V^2}{R}$ and in an A.C. circuit it is $\dfrac{V_{rms}^{2}}{R}$.

          ....

          ....

          .... **(3 marks)**

     (ii) Use the axes below to sketch a graph of how the power transferred to the
          lamp varies over one cycle. There is no need to put values on the power axis
          but you should include values on the time axis.   **(3 marks)**

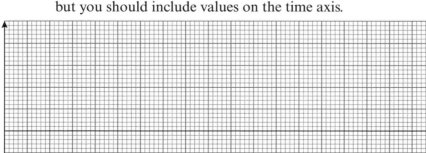

☷ **Maths skills** Power
depends
on voltage squared,
so can it be
negative?

Exam skills 9
Electromagnetic fields

1 The figure below shows how the current through a coil varies with time.

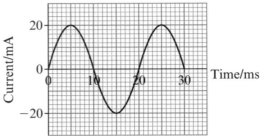

Maths skills

(a) Calculate the frequency of the A.C. current.

....

.... **(2 marks)**

(b) The resistance of the coil is $5.0\,\Omega$. Calculate the electrical power dissipated as heat in the coil.

....

....

.... **(3 marks)**

A second coil is placed close to the first one, as shown in the figure. Its terminals are X and Y. When an oscilloscope is connected to terminals X and Y, the trace on the screen shows an alternating e.m.f. between the terminals.

first coil   second coil   X Y

(c) Explain why an induced e.m.f. is produced across the second coil.

....

....

....

....

....

.... **(4 marks)**

(d) Discuss whether there is any energy transfer between the two coils.

....

....

....

.... **(2 marks)**

The Rutherford scattering experiment

1   Rutherford's alpha-particle scattering experiment provided evidence for the nuclear model of the atom.

(a)   Why was the experiment carried out in a vacuum?

.... **(1 mark)**

(b)   Why was gold foil used as a target?

.... **(1 mark)**

(c)   Which piece of evidence suggested that the atom is mainly empty space?

....

....

.... **(2 marks)**

(d)   Explain how the scattering of alpha particles suggested that:

    (i)   the nucleus is charged.

....

.... **(2 marks)**

    (ii) the nucleus is tiny compared with the atom.

....

.... **(2 marks)**

(e)   Complete the diagram below to show the paths of the three alpha particles as they approach the gold nucleus.

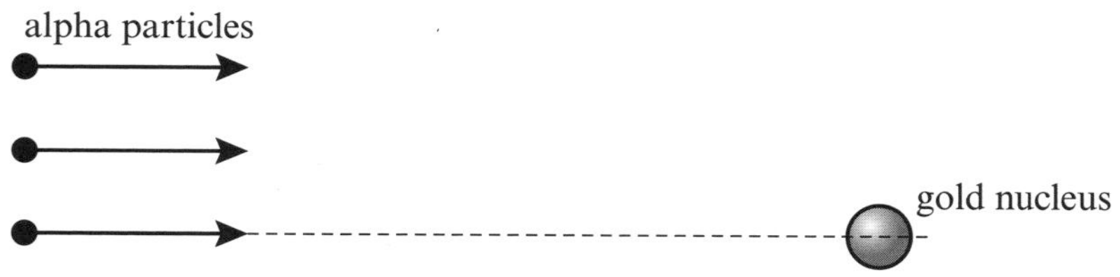

alpha particles

gold nucleus

(3 marks)

(f)   What does the Rutherford model suggest about the density of nuclear matter compared with the density of normal matter?

> Think about the volume of the atom and the mass of the particles in it according to the Rutherford model.

....

....

(2 marks)

> Guided

(g)   Gold has atomic number 79. Tungsten has atomic number 74. Suggest how the results of a similar experiment carried out with a very thin layer of tungsten would differ from those carried out using a gold foil.

With a lower atomic mass there will be a lower positive charge on the nucleus of

tungsten and so it will exert a smaller force ....

....

.... **(3 marks)**

Had a go ☐   **Nearly there** ☐   **Nailed it!** ☐

Nuclear notation

⟩**Guided**⟩ **1** (a) Complete the table below to show the numbers of protons, neutrons and electrons in neutral atoms of each nuclide.

Nuclide	Number of protons	Number of neutrons	Number of electrons
$^{7}_{3}\text{Li}$	3	$7 - 3 = 4$	3
$^{108}_{47}\text{Ag}$			
$^{222}_{86}\text{Rn}$			
$^{244}_{94}\text{Pu}$			

(12 marks)

(b) Radon-222 decays to polonium by emitting an alpha particle.

(i) State the structure and charge of an alpha particle.

....

....

.... **(3 marks)**

(ii) Write a balanced nuclear equation for the alpha decay of radon-222.

.... **(3 marks)**

(iii) Explain how nucleon number and charge are conserved in the alpha decay in (ii).

....

.... **(2 marks)**

2 A model of the atom that was proposed at the start of the twentieth century is referred to as the 'plum pudding' model because it assumes that the atom is a sphere of positively charged matter with negatively charged electrons embedded within it and held by electrostatic forces. Explain how this model was contradicted by the results of Rutherford's scattering experiment.

....

....

....

....

┌──────────────────────────────────────┐
│ It is easy to stray from the question │
│ so read it carefully! You need │
│ to start with evidence from the │
│ scattering experiment and then │
│ show how this cannot be explained │
│ by the 'plum pudding' model. │
└──────────────────────────────────────┘

....

....

....

.... **(4 marks)**

Electron guns and linear accelerators

1   The figure below shows an electron beam tube.

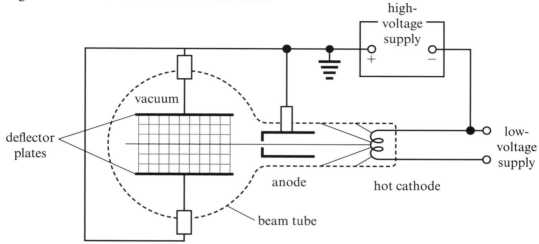

(a) (i)   Explain the purpose of the low-voltage supply.

....

.... **(2 marks)**

(ii) Explain the purpose of the high-voltage supply.

....

.... **(2 marks)**

(iii) Explain why it is essential for the beam tube to contain a vacuum.

....

.... **(2 marks)**

(b) In a particular experiment, the high-voltage supply is set to 2500 V.

(i)   Calculate the kinetic energy of electrons (in joules) as they pass the anode.
(The charge on an electron $e = -1.6 \times 10^{-19}$ C; $m_e = 9.1 \times 10^{-31}$ kg.)

....

.... **(2 marks)**

(ii) Calculate the velocity of the electrons as they pass the anode.

....

.... **(2 marks)**

(iii) For this tube, the electron beam will not be
deflected as it passes between the two 'deflector'
plates. Explain why not.

Look carefully at the diagram. What must be present for the beam to be deflected?

....

.... **(2 marks)**

Cyclotrons

1   A cyclotron used to accelerate electrons has a uniform magnetic field of strength B between the poles of its electromagnet. The radius of the cyclotron is r.

The potential difference between the two dees is V.

Electrons are injected into one of the dees near the centre of the cyclotron, as shown in the figure below.

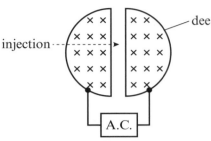

(a)   Explain why the electrons will follow a semicircular path in the dees.

....

....

.... **(2 marks)**

Guided   (b)   Show that the radius of curvature r of the path inside one of the dees is given by $r = \dfrac{mv}{Be}$ where m is the mass of the electron, e is its charge and v is its speed.

> The equation for centripetal force is $F = \dfrac{mv^2}{r}$.

$F = Bev =$ ....

....

.... **(2 marks)**

Guided   (c)   Show that the time taken for the electron to complete a semicircular path inside the dee does not depend on the speed or kinetic energy of the electron.

Time $= \dfrac{distance}{speed}$, so $T = \frac{1}{2} \times \dfrac{2\pi r}{v}$ ....

....

.... **(3 marks)**

(d)   Explain why the dees are connected to an alternating supply of constant frequency.

....

....

.... **(3 marks)**

(e)   Explain why the electrons will eventually leave the cyclotron.

....

.... **(2 marks)**

Particle detectors

1 A particular cyclotron has radius 34 cm and a magnetic field strength of 0.44 T. The peak A.C. voltage between the dees is 5000 V.

(a) Calculate the cyclotron frequency for this cyclotron.
    ($e = -1.6 \times 10^{-19}$ C; $m_e = 9.1 \times 10^{-31}$ kg.)

....

....

.... **(3 marks)**

(b) Calculate the maximum energy for electrons accelerated by this cyclotron.

....

....

| The electrons will have their maximum energy when the radius of their orbit is equal to the radius of the cyclotron. |

.... **(3 marks)**

2 (a) Explain why it is much harder to detect neutrons or neutrinos than alpha particles or protons.

....

.... **(2 marks)**

(b) The figure below shows the structure of a Geiger–Müller tube.

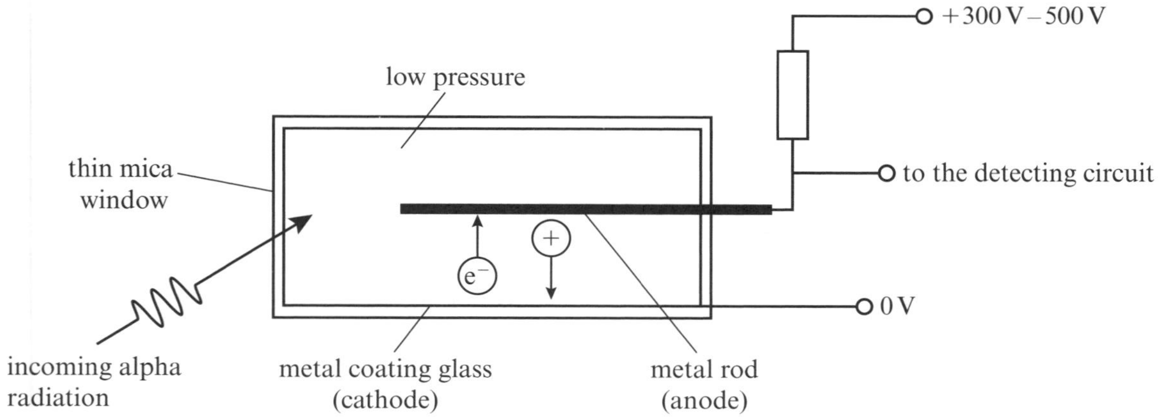

(i) What property of alpha radiation allows it to be detected by the Geiger–Müller tube?

.... **(1 mark)**

(ii) In order to detect alpha radiation a Geiger–Müller tube must have a very thin mica window. Explain why.

....

.... **(2 marks)**

Matter and antimatter

1 Which of the following statements about the positron is correct?

    ☐ **A** It has the same charge (magnitude and sign) as the electron.

    ☐ **B** It has the same mass as the electron.

    ☐ **C** It has the same lepton number (magnitude and sign) as the electron.

    ☐ **D** It has the same electric field (magnitude and sign) as the electron.   **(1 mark)**

2 Antimatter hydrogen atoms were created and trapped by physicists at CERN in 2010.

    (a) What particles make up an antihydrogen atom?

    ....   **(2 marks)**

Guided   (b) Calculate how much energy would be released if an antihydrogen atom annihilated with a hydrogen atom.

Mass of electron $m_e = 9.11 \times 10^{-31}$ kg; mass of proton $m_p = 1.67 \times 10^{-27}$ kg,

speed of light $c = 3.00 \times 10^8$ m s^{-1}.

Mass of a hydrogen atom is approximately ....

....

....

> Start by looking up the values of the constants you need in the data sheet.

    (4 marks)

    (c) Hydrogen bombs release energy from nuclear fusion, not matter–antimatter annihilation. A 1 megatonne hydrogen bomb releases about 4.2×10^{15} J. Calculate the total mass of hydrogen and antihydrogen needed to release the same amount of energy.

    ....

    ....   **(2 marks)**

3 (a) It is possible for a single gamma ray to create an electron–positron pair as it interacts with an atomic nucleus. It is also possible for an electron–positron pair to annihilate to create a pair of gamma-ray photons. Write balanced nuclear transformation equations for each process.

    Pair creation: ....

    Pair annihilation: ....   **(4 marks)**

    (b) (i) Calculate the maximum wavelength of each of the two gamma rays created when an electron and a positron annihilate. (Planck constant $= 6.63 \times 10^{-34}$ J s.)

    ....

    ....

> The maximum wavelength corresponds to the minimum energy.

    ....   **(3 marks)**

      (ii) Explain why this is a maximum wavelength.

    ....

    ....   **(2 marks)**

The structure of nucleons

1 In an experiment to measure the diameter of a proton, a high-energy electron beam was scattered from a liquid hydrogen target.

(a) Explain why liquid hydrogen was a suitable target material.

....

.... **(2 marks)**

> **Guided**

(b) (i) Explain why electrons must be accelerated to very high energy if they are to be used to measure the size of a proton.

> Think about the relationship between momentum and de Broglie wavelength.

In order to resolve detail, the de Broglie wavelength λ of the electron must be

....

....

.... **(2 marks)**

(ii) In this experiment, the electron momentum was $2.1 \times 10^{-18}\,\text{kg m s}^{-1}$. Estimate the radius of the proton.

....

.... **(2 marks)**

2 Baryons are formed from quark triplets. The proton is uud and the neutron udd. There are other closely related baryons, such as delta particles.

(a) The quark compositions of two of the delta (Δ) particles are shown below. State the charge on each particle.

Δ_1 (uuu) ....

Δ_2 (ddd) .... **(2 marks)**

(b) Antiquarks also combine in triplets to form antiparticles. What is the charge on the antiparticle $\bar{u}\,\bar{u}\,\bar{d}$?

.... **(1 mark)**

Nuclear energy units

1 The maximum energy of the Large Hadron Collider at CERN is 14 TeV (14×10^{12} eV).

(a) State what is meant by **one electronvolt** and give its value in joules.

....

....

.... **(2 marks)**

(b) Convert 14 TeV to joules.

....

.... **(2 marks)**

⟩**Guided**⟩ (c) Imagine that all the energy of one 14 TeV collision could be transferred to kinetic energy of a grain of sand of mass 0.005 g. Calculate the velocity of the grain of sand.

$KE = \frac{1}{2} mv^2 = 2.24 \times 10^{-6}$ J, so ....

$v =$ ....

.... **(2 marks)**

2 Show that $\dfrac{MeV}{c^2}$ has units of mass.

....

.... **(2 marks)**

🖩 **Maths skills** **3** A muon is a subatomic particle very similar to an electron but with a different mass. The mass of an electron is 9.11×10^{-31} kg and the mass of a muon is $106 \dfrac{MeV}{c^2}$.

(a) Convert the mass of the electron to $\dfrac{MeV}{c^2}$.

....

.... **(1 mark)**

(b) Calculate the ratio of the muon mass to the electron mass.

.... **(1 mark)**

(c) Calculate the rest energy of the electron in both electronvolts and joules.

....

....

.... **(2 marks)**

> If the rest mass of a particle is $1 \dfrac{MeV}{c^2}$ its rest energy is 1 MeV.

The Standard Model

1 Carbon-14 nuclei are unstable and decay by beta-minus decay.
   Here is an equation for the decay:

   $$^{14}_{6}\text{C} \rightarrow \, ^{14}_{7}\text{N} + \, ^{0}_{-1}e + \, ^{0}_{0}\nu$$

   (a) Explain why the emission of a beta particle also requires the emission of an antineutrino.

   ....

   ....

   ....

   > Think about conservation of lepton number and the fact that the antineutrino is an antiparticle.

   (2 marks)

> **Guided**

   (b) How does the nitrogen nucleus differ from the carbon nucleus in this decay?

   The proton number Z ....

   The neutron number N ....

   .... **(2 marks)**

   (c) It is suggested that the underlying decay is actually:

   $$^{1}_{0}\text{n} \rightarrow \, ^{1}_{1}\text{p} + \, ^{0}_{-1}e + \, ^{0}_{0}\bar{\nu}$$

   (i) Complete the table below by ticking the correct box to identify each particle by type:

	Quark	Lepton	Meson	Baryon
proton				
neutron				
electron				
antineutrino				

   (4 marks)

   (ii) Explain how baryon number is conserved in beta decay.

   .... **(1 mark)**

2 Particle physicists often carry out experiments in which proton beams collide with a target containing more protons (e.g. liquid hydrogen). This often results in the creation of pions. Here are two possible reactions in which pions are created:

   $$^{1}_{1}\text{p} + \, ^{1}_{1}\text{p} \rightarrow \, ^{1}_{1}\text{p} + \, ^{1}_{1}\text{p} + \, ^{0}_{0}\pi^{0}$$

   $$^{1}_{1}\text{p} + \, ^{1}_{1}\text{p} \rightarrow \, ^{1}_{1}\text{p} + \, ^{1}_{0}\text{n} + \, ^{0}_{1}\pi$$

   (a) What type of particle are the pions?

   ....

   (b) What is the baryon number of a pion?

   ....

   (c) What is the structure of a π^{0} particle?

   ....

   (d) What is the structure of a π^{+} particle?

   .... **(4 marks)**

Particle interactions

1 Here is an equation for the reaction that occurs when a pi-minus meson interacts with a proton to produce a sigma particle. Use conservation laws to work out the lepton number, baryon number and charge of the sigma particle.

$$^{0}_{-1}\pi^{-} + {}^{1}_{1}\text{p} \rightarrow {}^{0}_{-1}\pi^{-} + \sum$$

Lepton number: ....

Baryon number: ....

Charge: .... **(3 marks)**

2 Use conservation laws to work out the lepton number, baryon number and charge on the lambda particle that decays to form a proton and a pi-minus particle.

$$\Lambda \rightarrow {}^{1}_{1}\text{p} + {}^{0}_{-1}\pi^{-}$$

Lepton number: ....

Baryon number: ....

Charge: .... **(3 marks)**

3 A student suggests that the delta-minus particle (ddd) should be able to decay to an electron and a pi-zero as shown in the equation below:

$$\Delta^{-} \rightarrow \text{e}^{-} + \pi^{0}$$

Discuss, with reference to the relevant conservation laws, whether or not this decay is possible.

> Consider each conservation law in turn and see whether it holds for the reaction.

....

....

.... **(3 marks)**

4 The delta-2+ particle (uuu) can decay by the mechanism shown in the equation below:

$$^{1}_{2+}\Delta^{++} \rightarrow {}^{1}_{1}\text{p} + {}^{0}_{1}\pi^{+}$$

The mass of the proton is m, the mass of the delta-2+ is $1.2m$, and the mass of the pi-plus particle is $0.14m$.

(a) Find an expression for the energy released in the decay of a stationary delta-2+ particle in terms of m and c (the speed of light).

....

.... **(2 marks)**

(b) Show that the velocities of the proton and the pi-plus particle after the decay will be in the ratio of approximately 1 to 7.

....

....

> Use conservation of linear momentum.

.... **(3 marks)**

Exam skills 10
Nuclear and particle physics

1 The figure below shows a particle of mass *m* and charge +*q* entering a region of magnetic field of flux density *B* that is perpendicular to its velocity.

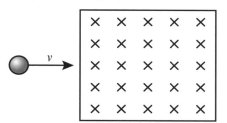

magnetic field into plane of
paper of flux density *B*

(a) Explain why the charged particle will move in an arc of a circle, and derive an expression for the radius of this circular path.

....

....

.... **(4 marks)**

(b) The figure shows a simplified diagram of a cyclotron used to accelerate protons.
A radio-frequency voltage supply is connected to the two dees.
The frequency of the supply is *f* and the potential difference between the dees when the protons cross the gap between them is *V*.

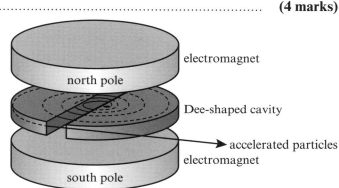

electromagnet

north pole

Dee-shaped cavity

accelerated particles

electromagnet

south pole

  (i) Derive an expression for the cyclotron frequency and show that it is independent of the speed of the protons. (Treat the size of the gap between the dees as negligible.)

....

....

.... **(4 marks)**

> **Guided**

  (ii) The diameter of this cyclotron is *D*. Derive an expression for the maximum achievable proton energy from this cyclotron.

KE of proton is $\frac{1}{2}mv^2 = $ ....

....

when $r = \frac{D}{2}$, KE is maximum ....

.... **(4 marks)**

  (iii) Calculate the maximum energy of protons accelerated in a cyclotron of diameter 1.2 m with a uniform magnetic field of flux density 0.80 T and express your answer in joules and electronvolts.

....

....

.... **(3 marks)**

Specific heat capacity

1   A block of ice of mass 2.5 kg is at a temperature of $-10\,°C$. How much energy must be supplied to raise the temperature of the ice to its melting point? (The specific heat capacity of the ice is $2100\,J\,kg^{-1}\,°C^{-1}$.)

....

.... **(2 marks)**

2   Describe an experiment to measure the specific heat capacity of a non-flammable liquid, such as water. Include a labelled diagram and explain how you would use the measurements taken to calculate a value for the specific heat capacity.
Explain any precautions you would take to ensure an accurate result.

....

....

....

> **Practical skills**   Think of ways to reduce heat losses in order to ensure a more accurate result.

....

....

....

....

.... **(8 marks)**

Guided

3   A block of copper of mass 250 g and at a temperature of $50\,°C$ is submerged in $0.0020\,m^3$ of water at a temperature of $20\,°C$. Calculate the final temperature of the water and the block, assuming that there are no heat losses to the surroundings. (Specific heat capacity of copper = $385\,J\,kg^{-1}\,°C^{-1}$; specific heat capacity of water = $4200\,J\,kg^{-1}\,°C^{-1}$; density of water = $1000\,kg\,m^{-3}$.)

energy transferred from copper to water $E = m_{Cu}c_{Cu}\Delta\theta_{Cu} = m_{W}c_{W}\Delta\theta_{W}$

....

....

....

> Equate the heat lost by the copper to the heat gained by the water.

if final temperature of water and copper = T, then $\Delta\theta_{Cu} = 50 - T$ and

$\Delta\theta_{W} = T - 20$ ....

....

.... **(4 marks)**

Latent heats

1   (a)   Explain why adding an ice cube at 0 °C to a cold drink will keep it cold for longer than adding the same mass of water at 0 °C.

....

....

.... **(3 marks)**

   (b)   Ice cubes of total mass 50 g are added to a flask containing 500 cm³ of orange squash, initially at 15.0 °C. The specific heat capacity of the orange squash can be taken to be the same as water (4200 J kg⁻¹ °C⁻¹) and the latent heat of fusion for water is 334 kJ kg⁻¹.

     (i)   Calculate the energy needed to completely melt the ice cubes at their melting point.

....

.... **(2 marks)**

     (ii)   Calculate the temperature drop of the orange squash if it supplies the energy to melt the ice in (b)(i).

....

.... **(2 marks)**

Practical skills

Guided

2   An early method for finding the specific latent heat of vaporisation of a liquid used an apparatus like the one shown in the figure.

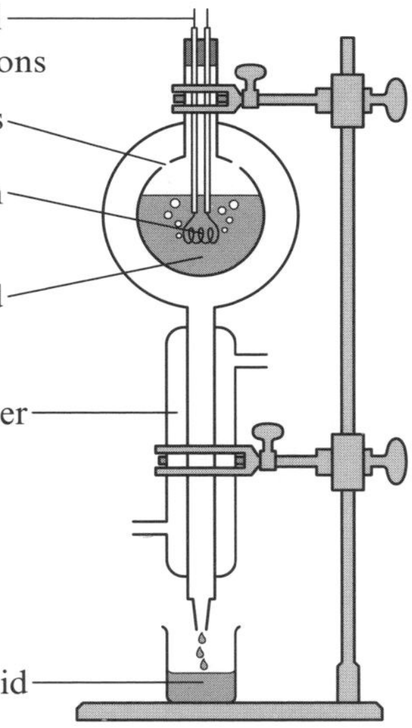

   (a)   Explain how this apparatus was used to find the latent heat of vaporisation of a liquid. Your answer should include a list of the measurements to be made and should explain how the results are used to calculate a value for the latent heat of vaporisation.

electrical connections

holes

platinum heater

liquid

condenser

liquid

Start by explaining how the apparatus works:

electrical energy supplied to the heater is

used to boil the liquid, ... Then state the

measurements needed: mass, ...

....

....

....

....

....

Finally describe how the latent heat can be calculated.

(6 marks)

   (b)   Explain why it would be important to wait until the liquid was boiling before beginning to measure the electrical energy supplied to the heater.

....

.... **(2 marks)**

Had a go ☐ Nearly there ☐ Nailed it! ☐

Pressure and volume of an ideal gas

Practical skills

1 The figure shows some apparatus that can be used to measure how the volume of a gas depends on its pressure. Pressure is applied to the oil in the reservoir, forcing it into the tube and compressing the air trapped in the tube. The length of the air column in the tube can be measured against the scale.

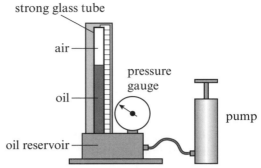

(a) Explain why it is important for the air column to be trapped in a tube of constant cross-sectional area.

....

.... **(2 marks)**

(b) Explain why it is important to compress the gas *slowly*.

....

.... **(2 marks)**

(c) Boyle's law suggests that the volume of a constant mass of an ideal gas at constant temperature is inversely proportional to the pressure of the gas. Explain how results taken from an experiment like the one shown above can be used to test this relationship.

> What graph could be plotted? Simply plotting volume (or length of air column) against pressure will not be good enough.

....

....

....

....

.... **(4 marks)**

Guided

2 (a) The pressure inside an oxygen cylinder is 1.2×10^7 Pa. The cylinder has a volume of $0.0040\,m^3$. The oxygen is released into the atmosphere. Calculate the volume occupied by the oxygen from the cylinder when it is at atmospheric pressure (1.0×10^5 Pa).

$p_1V_1 = p_2V_2$ ....

....

.... **(2 marks)**

(b) State two assumptions that you made in carrying out your calculation in part (a) above.

....

.... **(2 marks)**

Absolute zero

1 Which of the following statements about temperature is incorrect?

   ☐ **A** A temperature rise of 10 °C is the same as a temperature rise of 10 K.

   ☐ **B** The temperature of snow is around 273 K.

   ☐ **C** The boiling point of nitrogen (−196 °C) is 77 K.

   ☐ **D** The melting point of lead (27 °C) is 54 K.   **(1 mark)**

2 State two pieces of experimental evidence that led to the idea that there is a lowest possible temperature.

   ....

   ....

   ....   **(2 marks)**

3 A flask contains 500 cm³ of air at room temperature 20 °C and a pressure of 102 kPa. The flask is heated to a temperature of 80 °C and the pressure increases to 125 kPa. Use these results to calculate an experimental value for the absolute zero of temperature. Assume that the pressure increases linearly with temperature.

   ....
   .... | Calculate the change in temperature corresponding to a change of 1 kPa in pressure and then work backwards to zero pressure. |
   ....
   ....

   ....   **(3 marks)**

4 (a) Convert the boiling point of oxygen from °C to K: −183 °C

   ....   **(1 mark)**

   (b) Convert the temperature at the surface of Venus from K to °C: 735 K

   ....   **(1 mark)**

5 Measurements of the variation of gas pressure with temperature for an ideal gas can be extrapolated back to find a value for the absolute zero of temperature. If a real gas were cooled down toward absolute zero, suggest with reasons how the variation of gas pressure with temperature might differ from the linear relationship extrapolated for an ideal gas.

   ....

   ....

   ....

   ....   **(3 marks)**

Kinetic theory

1   Use the kinetic theory model to explain why:

(a) Gases exert a pressure on the walls of their containers.

....

....

.... **(3 marks)**

(b) The pressure of a gas increases when it is compressed at constant temperature.

....

....

.... **(3 marks)**

(c) The pressure of a gas increases when it is heated at constant volume.

....

....

.... **(3 marks)**

2   The air in a child's balloon is at a temperature of 25 °C and a pressure of 110 kPa. The volume of the balloon is $3.0 \times 10^{-3}\,\text{m}^3$.

(a) Calculate the number of molecules of air inside the balloon.

....

....

....

.... **(3 marks)**

⟩**Guided**⟩   (b) Calculate the total internal energy of the gas in the balloon. (Assume that the internal energy of the gas is equal to the total kinetic energy of its molecules).

For one molecule, $\frac{1}{2}m\langle c^2 \rangle = \frac{3}{2}kT$ ....

....

.... **(3 marks)**

3   When tiny particles of smoke suspended in the air are observed under a microscope they are always moving in an erratic random motion, continually changing direction. This is called Brownian motion.
Use the kinetic theory to suggest an explanation for Brownian motion.

....

....

.... **(3 marks)**

Particles and energy

1 The internal energy of an ideal gas is doubled. How does this affect the mean kinetic energy of the molecules in the gas?

☐ **A** It does not affect them.

☐ **B** Their mean kinetic energy increases by a factor of $\sqrt{2}$.

☐ **C** Their mean kinetic energy increases by a factor of 2.

☐ **D** Their mean kinetic energy increases by a factor of 4.   **(1 mark)**

2 Six molecules in a gas have speeds: 406, 438, 395, 450, 448, 508 ms^{-1}.

(a) Calculate the mean speed $<c>$.

....   **(1 mark)**

(b) Calculate the mean square speed $<c^2>$.

....   **(1 mark)**

(c) Calculate the root mean square speed c_{rms}.

....   **(1 mark)**

3 (a) Describe the distribution of molecular energies in a gas at a given temperature.

....

....

....   **(3 marks)**

(b) The graph shows the distribution of particle energies in a gas at two different temperatures, T_1 and T_2. State, with reasons, which of T_1 and T_2 is the higher temperature.

....

....

....   **(2 marks)**

4 A hot air balloon contains air at a temperature of 40 °C.

(a) Calculate the mean kinetic energy of molecules in this balloon.

....

....   **(2 marks)**

(b) Calculate the root mean square speed of an air molecule in this balloon (take the mass of an air molecule to be approximately 4.7×10^{-26} kg).

....

....   **(2 marks)**

Black body radiation

1 The figure shows the spectrum of cosmic background radiation measured by the COBE satellite.

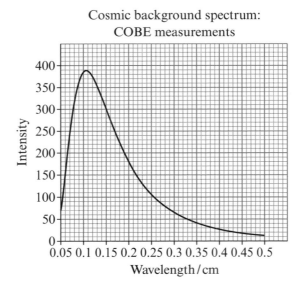

Cosmic background spectrum: COBE measurements

(a) State the wavelength at which there is a maximum intensity of background radiation.

.... **(1 mark)**

(b) The radiation is a close fit to a black-body radiation curve. Use your value from (a) to calculate the temperature of empty space.

....

....

You will need the constant in Wien's law, 2.898×10^{-3} m K.

(2 marks)

(c) To what part of the electromagnetic spectrum does this radiation belong?

.... **(1 mark)**

2 The star Betelgeuse has a surface temperature of 3500 K and a radius of 8.2×10^{11} m. Sirius has a surface temperature of 9940 K and a radius of 1.2×10^9 m.

(a) Sketch, on the axes, the radiation spectra from Betelgeuse and Sirius. Include approximate values and units on the wavelength axis. Do not put values on the intensity axis.

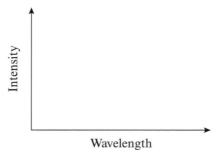

(5 marks)

(b) Calculate the total power radiated by Betelgeuse (its luminosity). (Stefan's constant, $\sigma = 5.67 \times 10^{-8}$ W m^{-2} K^{-4}.)

....

....

....

....

.... **(2 marks)**

Standard candles

1 The luminosity of star A is four times greater than the luminosity of star B but both stars appear equally bright from the Earth. If A is 80 light years away, how far away is B?

   ☐ **A** 5 light years   ☐ **B** 20 light years

   ☐ **C** 40 light years   ☐ **D** 160 light years   **(1 mark)**

2 Type 1a supernovae are extremely luminous and astronomers can predict their absolute luminosity. They are used as standard candles for distance measurements in astronomy.

   (a) Explain the advantage of high luminosity.

   ....

   .... **(2 marks)**

   (b) Explain why astronomers must be able to predict the absolute luminosity of a standard candle.

   ....

   .... **(2 marks)**

   (c) Describe how astronomers can use a standard candle to calculate the distance to a distant galaxy.

   ....

   ....

   .... **(3 marks)**

⟩Guided⟩   **3** The luminosity of the Sun is 3.8×10^{26} W and its distance from the Earth is 1.5×10^{11} m. The Hubble Space Telescope orbits the Earth and generates electricity from two solar panels of total area 35 m^2.

   (a) Calculate the intensity of solar radiation at the distance of the Earth from the Sun.

   $I = \dfrac{L}{4\pi d^2} = $ ....

   ....

   .... **(2 marks)**

   (b) Calculate the maximum power output from the solar panels, assuming that they have an efficiency of 20% and that the distance of the telescope from Earth is negligible in comparison with the distance to the Sun.

   Power = area × intensity × efficiency = ...

   ....

   .... **(3 marks)**

Trigonometric parallax

1   Long before Copernicus, the ancient Greek astronomer Aristarchus suggested that the planets, including the Earth, actually orbit the Sun. If this is the case, then the apparent positions of stars should shift as the Earth completes its orbit. However, these parallax effects were not observed until 1838.

   (a) Suggest a possible reason why the parallax effects were not observed until relatively recently.

   ....

   ....   **(1 mark)**

   (b) How does the size of these parallax movements depend on the distance from the Earth to the star?

   ....

   ....   **(1 mark)**

   (c) Earth-based telescopes can measure parallax angles for stars out to a distance of about 100 light years. However, the Hipparcos satellite can measure the parallax angles of stars out to about 1000 light years. Approximately how many times more stellar distances can be measured by Hipparcos than by ground-based telescopes? State any assumption that you make to answer this question.

   ....

   .... | **Maths skills** | The volume of a sphere is $\frac{4}{3}\pi r^3$ |

   ....

   ....   **(3 marks)**

2   The nearest star to Earth, apart from the Sun, is Proxima Centauri, which is 4.24 light years away. The Earth's orbital radius is 1.5×10^{11} m.

   | **Maths skills** | A sketch will help you visualise the trigonometry. |

Guided

   (a) Calculate the parallax angle for Proxima Centauri in degrees.

   1 light year = $3.00 \times 10^8 \times 3600 \times 24 \times 365$ ....

   $\tan \alpha = $ ....

   ....   **(3 marks)**

   (b) Explain why a long baseline is needed for stellar parallax measurements.

   ....

   ....   **(2 marks)**

The Hertzsprung–Russell diagram

1 (a) Sketch the Hertzsprung–Russell diagram, showing its four main regions.

(6 marks)

(b) Mark the position of the Sun on your diagram.   **(1 mark)**

(c) Name the process that fuels stars in the main sequence.

....   **(1 mark)**

2 Explain what is meant by each of the following:

(a) Stellar luminosity

....

....

....

> There are two marks per item, so try to make two points in each of your answers.

(2 marks)

(b) White dwarf star

....

....

....   **(2 marks)**

(c) Neutron star

....

....

....   **(2 marks)**

(d) Black hole

....

....

....   **(2 marks)**

(e) Main-sequence star

....

....

....   **(2 marks)**

Stellar life cycles and the Hertzsprung–Russell diagram

1 (a) State the property that determines the final stage in the life cycle of a star.

.... **(1 mark)**

   (b) Sketch a diagram to show the possible life cycles of stars, starting from a stellar nebula, as shown.

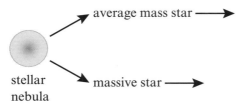

(9 marks)

2 The Hertzsprung–Russell diagram below shows the life cycle of a star like our Sun.

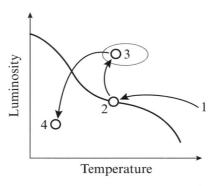

Temperature

Guided   (a) Describe the changes that take place through stages 1 to 4.

Stage 1: gas cloud collapses to form ....

....

Stage 2: ....

....

Stage 3: ....

....

Stage 4: ....

.... **(6 marks)**

   (b) What are the most significant differences between the life of a star like our Sun and the life of a much more massive star?

....

....

....

.... **(4 marks)**

The Doppler effect

1 A Formula 1 car races past a spectator at constant speed *v*, as shown in the figure below. The spectator notices that the engine note (the frequency of the sound) seems to change as the car passes.

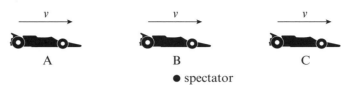

(a) Explain why the frequency of the engine note heard by the spectator changes.

....

.... **(2 marks)**

(b) Explain why the engine note heard by the driver of the car does not change as the car moves from A to C.

....

.... **(2 marks)**

(c) (i) On the axes below, sketch how the engine note heard by the spectator changes as the car moves from A to C. The dotted line represents the note heard by the driver.

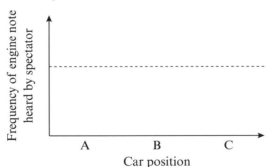

Think of the car's velocity as a vector. Does it have a component toward or away from the spectator?

(3 marks)

(ii) Explain the shape of your graph.

....

....

Explain each section: A to B, B, B to C.

....

.... **(3 marks)**

 2 Radio telescopes can detect radio waves from stars in the spiral arms of our galaxy. These have a characteristic wavelength of 21.106 cm if they are received from a stationary source. In one particular observation, the waves received on Earth are found to have a wavelength of 21.132 cm.

What can you conclude about the motion of these stars? Include a calculation.

The wavelength is redshifted, so ....

Doppler shift $z = $ .... **(2 marks)**

Cosmology

Guided 1 (a) Explain how the redshift of light from a distant galaxy can be used to estimate the distance to that galaxy. State any other pieces of information you would need in order to make the estimate.

Redshift, $z = \dfrac{\Delta\lambda}{\lambda_0}$, is related to recession velocity v by the equation $z = \dfrac{v}{c}$.

Recession velocity is related to distance by Hubble's law, $v = H_0 d$, so ....

....

....

.... **(4 marks)**

(b) Suggest two reasons why the result of your calculation in (a) would be an estimate rather than an accurate value.

....

....

.... **(2 marks)**

2 Explain how the following pieces of evidence support the theory that the Universe began in a hot Big Bang.

(a) The Universe is filled with microwave background radiation that has a black-body spectrum corresponding to a temperature of about 2.7 K.

....

....

....

> Make sure your answer refers to both the microwave background radiation and its temperature.

.... **(3 marks)**

(b) The spectra of light from stars in distant galaxies are all redshifted.

....

....

....

.... **(3 marks)**

3 (a) State Hubble's law as an equation and define the terms in it.

....

....

.... **(3 marks)**

(b) The current best value for the age of the Universe is 13.7 billion years.
Use this value to estimate the Hubble constant, including an appropriate unit.

....

....

.... **(2 marks)**

Exam skills 11 Thermal physics

1   (a)   Describe an experiment that can be used to investigate how the pressure of a constant volume and mass of gas varies with temperature. Your answer should include (i) a labelled diagram of the apparatus, (ii) an explanation of how the experiment is carried out and what measurements must be taken.

(i)

(3 marks)

(ii) ....

....

....

....

.... **(5 marks)**

(b)   Explain how the results of such an experiment can be used to find a value in degrees Celsius for the absolute zero of temperature.

....

....

.... **(2 marks)**

(c)   How is the existence of an absolute zero of temperature explained in terms of the kinetic theory?

....

....

.... **(2 marks)**

(d)   A small amount of bromine gas (Br_2) escapes into the air in a laboratory. The air temperature is 22 °C. The mass of a bromine molecule is 2.7×10^{-25} kg.

(i)   How does the mean kinetic energy of the bromine molecules compare with the mean kinetic energy of the air molecules? (Do not include a calculation.)

....

.... **(2 marks)**

 (ii)   Calculate the r.m.s. speed of these bromine molecules in air.

....

....

.... **(3 marks)**

Mass and energy

1   It is usually assumed that mass is conserved in chemical reactions even if energy is released. For example, when hydrogen is burnt in oxygen, the energy released is $1.43 \times 10^8\,\mathrm{J\,kg^{-1}}$.

Maths skills

   (a) (i) Calculate the mass equivalent to $1.43 \times 10^8\,\mathrm{J}$ of energy.

....

.... **(2 marks)**

       (ii) Hence, explain why it is sensible to assume that chemical reactions conserve mass.

....

.... **(2 marks)**

   (b) (i) Hydrogen is converted to helium by nuclear fusion reactions in the core of the Sun. The energy released is about $6.8 \times 10^{14}\,\mathrm{J\,kg^{-1}}$. Use this value to calculate the percentage change in mass in this nuclear fusion reaction.

....

....

.... **(2 marks)**

       (ii) Hence, state why it is important to consider mass changes in nuclear reactions.

.... **(1 mark)**

   (c) The most efficient way to release energy from matter is to combine matter and antimatter, when all mass is converted into energy. Using your answers to (a)(i) and (b)(i), calculate how many times greater the energy released per kilogram in matter–antimatter annihilation is than:

       (i) a chemical reaction, such as the combustion of hydrogen

....

.... **(2 marks)**

       (ii) a nuclear fusion reaction, such as the creation of helium in the core of the Sun.

....

.... **(2 marks)**

Guided   2   When an electron and a positron (antielectron) with negligible kinetic energy meet and annihilate, two gamma-ray photons of equal energy are emitted in opposite directions.

   Calculate the energy released when an electron and a positron annihilate in a low-energy collision. ($m_e = 9.1 \times 10^{-31}\,\mathrm{kg}$.)

   The total mass annihilated is the sum of the masses of the two particles = ...

   The energy released is given by the equation $E = mc^2$ = ... **(2 marks)**

Nuclear binding energy

1   Which of the following statements gives the most accurate definition of the binding energy of an atomic nucleus?

    ☐ **A**   The energy that binds the atom together.

    ☐ **B**   The energy stored in the nucleus.

    ☐ **C**   The energy needed to take the nucleus apart.

    ☐ **D**   The energy that binds the nucleons together.   **(1 mark)**

Guided

2   Calculate the binding energy per nucleon for a $^{16}_{8}O$ oxygen nucleus.
Express your answer in joules and electronvolts. (Mass of an oxygen nucleus 15.994915 u; $m_p = 1.007276$ u; $m_n = 1.008665$ u; 1 u $= 1.67 \times 10^{-27}$ kg.)

mass of 8 protons and 8 neutrons = ...

mass deficit = ...

total binding energy = ...

....

binding energy per nucleon = ...

....

> Work out the mass deficit for the nucleus compared with the nucleons and then convert this to energy. Don't forget to divide by the number of nucleons.

       (4 marks)

3   The figure shows how the binding energy per nucleon varies with mass number.

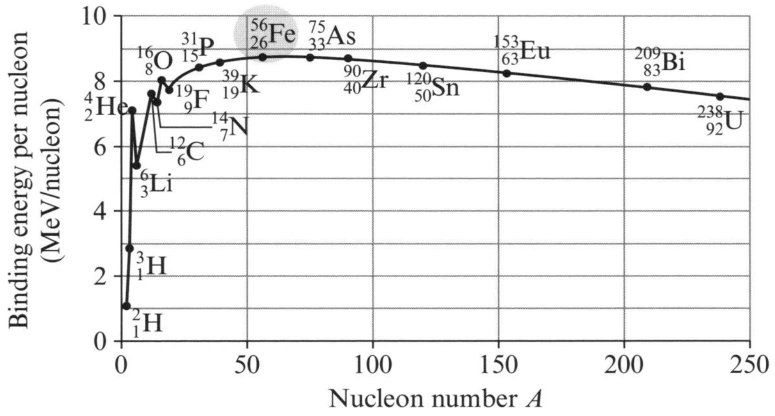

(a)   Explain why iron-56 is regarded as the most stable nuclide.

....

....   **(2 marks)**

(b)   After hydrogen the next three most abundant nuclides in the Milky Way are helium (He), oxygen (O) and carbon (C). Use the graph to explain why this might be.

....

....

....

....   **(3 marks)**

Nuclear fission

1 (a) In the space below, sketch a graph of nuclear binding energy against nucleon number.

(3 marks)

   (b) Explain with reference to your diagram from part (a) how nuclear fission of heavy nuclei can release a large amount of energy.

   ....

   ....

> You might find it helpful to add lines or labels to your graph.

   ....

   ....   **(3 marks)**

2 Here is an incomplete nuclear equation for the fission of a uranium-235 nucleus after it absorbs a neutron. Write the missing values below.

$$^a_b n + {}^{235}_{92}U \longrightarrow {}^c_d Ba + {}^{92}_{36}Kr + 3\,^a_b n$$

a = ... b = ... c = ... d = ...   **(4 marks)**

Maths skills

3 The energy released by nuclear fission of one uranium-235 nuclide is about 200 MeV.

   (a) Use this value to calculate the mass deficit (in atomic mass units, u) for one nuclear fission reaction. ($1\,u = 1.67 \times 10^{-27}$ kg.)

   ....

   ....

   ....   **(3 marks)**

Guided

   (b) Calculate the energy released if all of the atoms in 1.0 g of uranium-235 were to undergo nuclear fission.

   $N = \dfrac{1.0}{235} \times 6.02 \times 10^{23}$ ....

   ....

> First, calculate the number of atoms using the Avogadro number (6.02×10^{23}). Then multiply by the energy per fission in joules.

   ....

   Total energy = ...

   ....   **(3 marks)**

Nuclear fusion

1   (a)   State one similarity and one difference between nuclear fission and nuclear fusion.

....

.... **(2 marks)**

(b)   Describe the process of nuclear fusion and explain why it is hard to achieve in a controlled way.

....

....

....

.... **(4 marks)**

(c)   Two deuterium nuclei can fuse to form a tritium nucleus plus one proton. The equation for such a reaction is shown below.

$$_1^2H + _1^2H \rightarrow _1^3H + _1^1H$$

(i)   Show how this reaction conserves both charge and baryon number.

....

.... **(2 marks)**

(ii)   Calculate the energy released by this reaction. Express your answer in J and MeV. (Mass of a proton = 1.007276 u; mass of a deuterium nucleus = 2.013553 u; mass of a tritium nucleus = 3.015500 u; 1 u = 1.67×10^{-27} kg.)

.... | Start by working out the mass deficit for the reaction, then use $E = mc^2$ and finally convert units. |

....

....

....

.... **(5 marks)**

⟩Guided⟩   (iii)   Deuterium occurs naturally at a ratio of about 1 in 4500 hydrogen atoms. Suppose that all of the deuterium within 1.0 kg of seawater were to undergo nuclear fusion to form tritium by the reaction above. Use the ratio above and your answer to (c)(ii) to estimate the maximum energy that could be released from the seawater. Assume that the average molar mass of the water (H_2O) is 18 g and that dissolved salts can be neglected. (Avogadro number = 6.02×10^{23}.)

> Start by calculating the number of molecules of water and then find the number of deuterium atoms.

Number of hydrogen and deuterium nuclei in 1.0 kg water = ...

....

Number of deuterium nuclei only = ...

Maximum energy released = ...

.... **(3 marks)**

Background radiation

1 Commercial airline pilots and air crew experience higher average doses of background radiation than people who spend most of their lives at ground level. This is because:

☐ **A** the Earth absorbs cosmic radiation.   ☐ **B** the Earth is not radioactive.

☐ **C** the atmosphere absorbs cosmic radiation.   ☐ **D** the atmosphere is radioactive.   **(1 mark)**

2 In the 1950s and 1960s, nuclear weapons were tested in a large number of above-ground nuclear explosions. This resulted in radioactive contamination that increased the background count globally. Such tests were later banned, and the effect of this contamination on global average background counts peaked in 1963.

(a) Explain why nuclear explosions would increase the background count.

.... | Think about the products of
.... | nuclear fission reactions.   **(2 marks)**

(b) Explain why the effect of the nuclear contamination has been falling since 1963.

....

....   **(2 marks)**

3 People who are employed to work in a potentially radioactive environment, such as the nuclear power industry, are required to wear radiation monitoring badges. The figure shows the structure of such a badge.

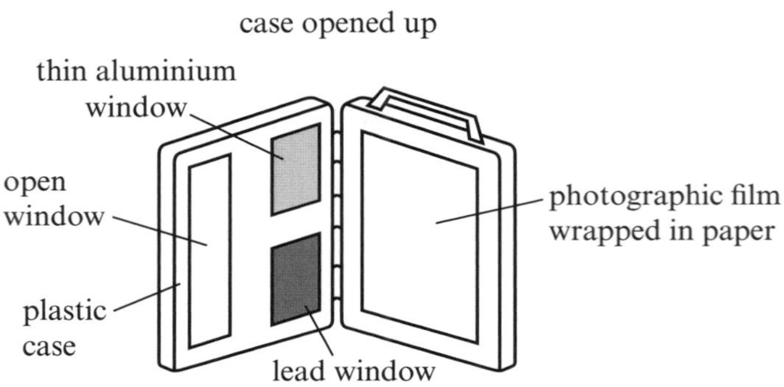

case opened up
thin aluminium window
open window
plastic case
lead window
photographic film wrapped in paper

The badge can be used to monitor both the total radiation dose and the types of radiation that the worker has been exposed to.

> **Guided**

Explain how exposure to different types of radiation (alpha, beta and gamma) can be inferred by comparing different parts of the film.

| Think about which type of
| radiation can pass through
| each window.

Each window ....

The open window ....

The thin aluminium window ....

The lead window ....

....   **(4 marks)**

Alpha, beta and gamma radiation

1 (a) Explain why gamma rays have a much longer range in air than beta particles or alpha particles.

....

.... **(2 marks)**

(b) Explain why alpha and beta particles can be deflected by electric and magnetic fields but gamma rays cannot.

....

.... **(2 marks)**

(c) Explain why alpha sources are far more dangerous to human beings if taken inside the body (e.g. inhaled radon gas) than if accidentally handled.

....

.... **(2 marks)**

2 A student is given a sample of a radioactive source to test. She uses a Geiger counter to measure count rates over 2-minute intervals with and without the source and with three different absorbers. Here are her results:

Source	Absorber	Counts (1)	Counts (2)	Counts (3)	Counts (4)	Counts (5)	Counts (avg)
none	none	22	20	20	25	18	21.0
present	none	77	83	85	81	80	....
present	thick card	72	74	78	75	80	....
present	2 mm Al	25	22	17	19	20	20.6
present	5 mm Pb	18	24	19	22	21	20.8

(a) Explain why repeat readings of the number of counts in any 2-minute interval vary.

....

.... **(1 mark)**

Guided (b) Complete the table by including the average counts. **(2 marks)**

(c) Identify the most likely type of radiation emitted by this source and explain your reasoning.

....

....

Use a process of elimination – which types of radiation can be stopped by each type of absorber?

....

.... **(4 marks)**

(d) State **three** safety precautions that should be observed by the student as she carries out this experiment.

....

....

.... **(3 marks)**

Investigating the absorption of gamma radiation by lead

1 A Geiger–Müller counter is used to record the intensity of radiation (the count rate) from a gamma-ray source in a school laboratory. The average count rate with no source present is 20 counts per minute (c.p.m.). The average count rate at a distance of 5.0 cm from the source is 68 c.p.m. What is the expected average count rate at a distance of 20.0 cm from the source?

☐ **A** 20 c.p.m.   ☐ **B** 23 c.p.m.

☐ **C** 32 c.p.m.   ☐ **D** 44 c.p.m.   **(1 mark)**

2 It is suggested that the intensity I of gamma radiation that passes through a lead sheet of thickness x is given by an equation of the form $I = I_0\, e^{-\mu x}$, where I_0 is the intensity of radiation incident on the sheet and μ is a constant called the absorption coefficient.

(a) State the S.I. units for μ.

.... **(1 mark)**

(b) A student decides to test the suggestion by measuring the intensity of gamma radiation that passes through lead sheets of various thickness.

   (i) State a suitable instrument with which to obtain a precise measurement of the thickness of the lead.

   .... **(1 mark)**

   (ii) Explain how the student can allow for the effects of background radiation.

   ....

   .... **(2 marks)**

   (iii) By taking natural logarithms of both sides of the equation above, show that a graph of $\ln I$ against x can be used to find the value of the absorption coefficient.

   ┌─────────────────────────────────┐
   │ **Maths skills** Remember that   │
   │ $\ln (e^{-\mu x}) = -\mu x$ and then │
   │ think about the   │
   │ equation of a straight   │
   │ line: $y = mx + c$.   │
   └─────────────────────────────────┘

   ....

   ....

   ....

   ....

   .... **(4 marks)**

Nuclear transformation equations

1 A student suggests that a neutron might decay by the process shown in this equation:

$$^{1}_{0}n \rightarrow {}^{1}_{1}p + {}^{0}_{-1}e$$

(a) Show that this reaction would conserve charge and baryon number.

....

.... (2 marks)

(b) Show that the reaction does not conserve lepton number and so cannot occur.

.... (1 mark)

(c) In fact, free neutrons can decay to protons if in the process they also emit an antineutrino. Write down a nuclear transformation equation for this decay and explain how it conserves lepton number.

....

....

.... (3 marks)

2 The figure below shows a naturally occurring decay chain that starts at uranium-238 and ends with the stable isotope lead-206.

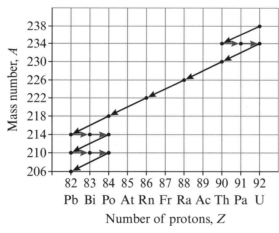

Think of the required change in baryon number (beta decays do not affect this).

(a) Write balanced nuclear equations for the decays of:

(i) thorium-234 .... (3 marks)

(ii) thorium-230 .... (3 marks)

◆Guided◆ (b) Another naturally occurring decay chain starts with thorium-232 and ends with lead-208.

(i) Determine how many alpha particles are emitted when one thorium-232 nucleus completes this decay chain.

Change in baryon number: 232 − 208 = ... (1 mark)

(ii) Determine how many beta particles are emitted when one thorium-232 nucleus completes this decay chain.

Protons in alpha particles emitted: .... (1 mark)

Radioactive decay and half-life

1   Two radioactive sources, X and Y, have equal activity at a particular time.
The half-life of X is 20 minutes and the half-life of Y is 30 minutes. Calculate the
ratio of the activity of X to the activity of Y after 1 hour.

☐ **A** 1:2   ☐ **B** 2:3   ☐ **C** 3:2   ☐ **D** 2:1   **(1 mark)**

2   In an experiment to determine the half-life of a certain mass of a radioactive gas,
an experimenter recorded the count rate per minute (c.p.m.) against time for a period
of 4 minutes. He then repeated the experiment with the same quantity of the same
radioactive gas. His results are shown in the table below:

Time /s	Trial 1 c.p.m.	Trial 2 c.p.m.	Mean c.p.m.
0	205	225	
30	155	163	
60	112	120	
90	83	93	
120	65	73	
150	50	58	
180	43	47	
210	36	40	
240	34	38	

(a) Complete the table by calculating the average count rate.   **(2 marks)**

(b) Use the graph paper below to plot a graph of average count rate against time.

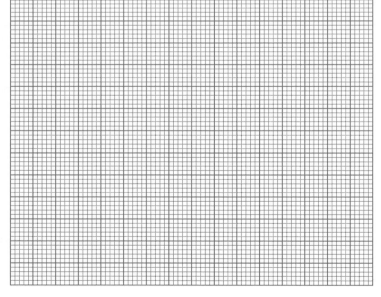

(3 marks)

(c) Use your graph to estimate the approximate average background count rate.

.... **(1 mark)**

(d) Use the graph to determine a value for
the half-life of this radioactive gas.
Show your working on the graph.

> Do not forget to take account of
> the average background count.

Half-life = ... **(3 marks)**

Exponential decay

1   In an experiment to determine the half-life of a certain mass of a radioactive gas, a scientist recorded the count rate per minute against time. She then repeated the experiment with the same quantity of the same radioactive gas. She calculated the mean count per minute and subtracted the background count per minute to get a corrected mean count per minute. The results are shown in the table below.

Time /s	Trial 1 c.p.m.	Trial 2 c.p.m.	Average c.p.m.	Corrected rate c.p.m.	ln(c.c.p.m.)
0	205	225	215	187	5.23
30	155	163	159	131	4.88
60	112	120	116	88	4.48
90	83	93	88	60	4.09
120	65	73	69	41	
150	50	58	54	26	
180	43	47	45	17	
210	35	43	39	11	
240	32	36	34	6	

(a)   Complete the table by calculating the natural logarithms ln(c.c.p.m.).   **(2 marks)**

Guided

(b)   Use the grid below to plot a graph of ln(c.c.p.m.) against time and use it to determine the half-life of the radioactive gas.

Maths skills   Take ln of both sides of:
$A = A_0 e^{-\lambda t}$
$\ln A = \ln A_0 - \lambda t$
Now compare this to
$y = mx + c$
What is the gradient of a graph of ln A vs t?

Decay constant λ is negative gradient of graph.

Half-life $t_{1/2} =$ ....

....   **(6 marks)**

Radioactive decay calculations

1 Cobalt-60 is produced in nuclear reactors. A laboratory uses a small sample of cobalt-60 as a gamma-ray source. The activity of the source when supplied new was 186 kBq, and the half-life of cobalt-60 is 5.3 years. The laboratory acquires a newly produced cobalt-60 source of the same mass.

Guided

(a) A student uses a Geiger–Müller tube to measure the intensity of gamma radiation from the newly acquired cobalt-60 source and finds that the count rate is very much less than 186 000 counts per second. Suggest three reasons for this.

The gamma rays are emitted randomly in all directions so ....

....

The efficiency of the detector is ....

The body of the sample itself ....

.... **(3 marks)**

(b) Calculate the activity of the old source:

   (i) 5.3 years after purchase

   .... **(1 mark)**

   (ii) 15.9 years after purchase.

   ....

   .... **(2 marks)**

(c) (i) Calculate the decay constant for cobalt-60 and state its unit.

   ....

   .... **(1 mark)**

   (ii) The source must be replaced when its activity falls below 10% of its initial activity. How often must cobalt-60 sources be replaced?

   ....

   ....

   ....

   .... **(3 marks)**

Gravitational fields

1 The mass of the Earth is about 75 times greater than the mass of the Moon. What is the ratio of the gravitational force of the Earth acting on the Moon to the gravitational force of the Moon acting on the Earth?

☐ **A** 1:75

☐ **B** 75:1

☐ **C** 1:1

☐ **D** 75^2:1

(1 mark)

Guided

2 During the Apollo 14 mission, astronaut Alan Shepard hit a golf ball and claimed that it travelled for 'miles and miles'. Given that a good golfer on Earth can drive a ball over 250 m and that the gravitational field strength g at the surface of the Earth is 9.81 N kg⁻¹ and at the surface of the Moon is 1.63 N kg⁻¹, is it possible that Alan Shepard's claim was true? Your answer should be supported by relevant calculations. (1 mile is approximately 1.6 km.)

The time of flight (time to fall to the ground) increases by a factor of $\dfrac{9.81}{1.63}$

....

....

....

....

....

....

> Assume that the ball is struck with the same initial speed and in the same direction as on Earth. How will the reduced value of g affect its time of flight? Are there any other relevant differences on the Moon?

(4 marks)

3 A moon rock is brought to Earth. Which of its properties will change as a result?

☐ **A** Mass

☐ **B** Weight

☐ **C** Density

☐ **D** Volume

(1 mark)

4 (a) State what is meant by a **uniform gravitational field**.

....

(1 mark)

(b) Under what circumstances can the gravitational field of the Earth be treated as uniform? (The radius of the Earth is 6400 km.)

....

(1 mark)

Gravitational potential and gravitational potential energy

1 The Hubble Space Telescope orbits at an altitude of 560 km above the Earth's surface.

  (a) Calculate the gravitational potential difference between the altitude of the Hubble Space Telescope and the surface of the Earth by assuming that the Earth's field is approximately uniform over this distance. (Assume $g = 9.81\,\text{N}\,\text{kg}^{-1}$.)

....

.... **(2 marks)**

  (b) Would the actual gravitational potential difference in (a) be greater than or less than the value you have calculated? Explain your answer.

....

....

.... **(2 marks)**

⟩**Guided**⟩ 2 A children's tower is constructed by stacking eight blocks one on top of the other. Each block is cubic and has a side length of 4.0 cm and a mass of 200 g. The tower falls down so that all of the blocks are resting on the ground. Calculate the loss of gravitational potential energy.

> Think about the amount by which the centre of mass drops.

gravitational potential energy of tower ....

....

gravitational potential energy of all bricks on floor ....

....

ΔGPE = ...

.... **(4 marks)**

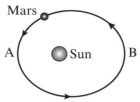

Newton's law of gravitation

1   The orbit of the planet Mars is elliptical and the planet's distance from the Sun varies. The figure shows the shape of its orbit (the ellipse is exaggerated). At their closest approach, the distance between the Sun and Mars is 207 million km. At their farthest point, the distance is 249 million km.

(a)   Calculate the ratio of the gravitational force on Mars at closest approach (position A) to the gravitational force on Mars at its farthest point from the Sun (position B).

....

.... **(2 marks)**

(b)   Add an arrow to the diagram to show the direction of the resultant force on Mars at the position shown.   **(1 mark)**

(c)   Describe the changing speed and path of Mars at the position shown.

.... **(1 mark)**

(d)   Describe how the gravitational potential energy and kinetic energy of Mars vary as it completes one orbit.

At B, Mars is at its farthest point from the Sun, so it has its greatest gravitational

potential energy. ....

....

.... **(4 marks)**

(e)   At which point in its orbit does Mars have its maximum speed?

.... **(1 mark)**

(f)   Mars has a mass of 6.39×10^{23} kg and the Sun has a mass of 1.99×10^{30} kg. Calculate the maximum gravitational attraction between Mars and the Sun.

....

.... **(2 marks)**

2   The Moon has a mass of 7.35×10^{22} kg and a radius of 1740 km.

(a)   Calculate the gravitational field strength at the surface of the Moon.

....

.... **(2 marks)**

(b)   Calculate the weight of a 65 kg woman on the surface of the Moon.

.... **(1 mark)**

121

Gravitational field of a point mass

1 On an imaginary line from the centre of the Earth to the centre of the Moon there will be a point at which the Moon's gravitational field strength is equal and opposite to the Earth's gravitational field strength. This is called the neutral point. The mass of the Earth is approximately 80 times larger than the mass of the Moon.

(a) Write down an expression for gravitational field strength g at a distance x from a mass m.

.... **(1 mark)**

(b) (i) The neutral point is at distance a from the Earth and b from the Moon. Calculate the ratio $\frac{a}{b}$.

....

....

.... **(3 marks)**

(ii) The distance from the Earth to the Moon is 3.84×10^8 m. Calculate distance a.

....

.... **(2 marks)**

(c) Discuss the significance of the neutral point for sending spacecraft from the Earth to the Moon.

....

.... **(2 marks)**

2 (a) Explain why the gravitational field strength of a planet decreases as you go beneath the surface toward the centre.

....

....

....

> Imagine that a point inside the surface rests on the surface of a smaller sphere of matter. How will the inner sphere and outer shell affect a mass placed at that point?

(2 marks)

(b) State and explain the value of the gravitational field strength at the centre of the Earth.

....

.... **(2 marks)**

Gravitational potential in a radial field

1 (a) Explain what is meant by 'the gravitational potential at a point in space'.

....

.... **(2 marks)**

   (b) Explain why the gravitational potential at any point in space is negative.

....

....
| Think about the work that must be
| done to move a mass to infinity.

(2 marks)

   (c) State the connection between gravitational field strength and gravitational potential.

....

.... **(2 marks)**

   (d) Explain how changes of gravitational potential energy are related to gravitational potential.

....

.... **(2 marks)**

2 The figure below shows equipotentials close to the surface of the Earth. These mark areas in which an object would have the same gravitational potential energy. The vertical distance between adjacent equipotentials is 1.0 m.

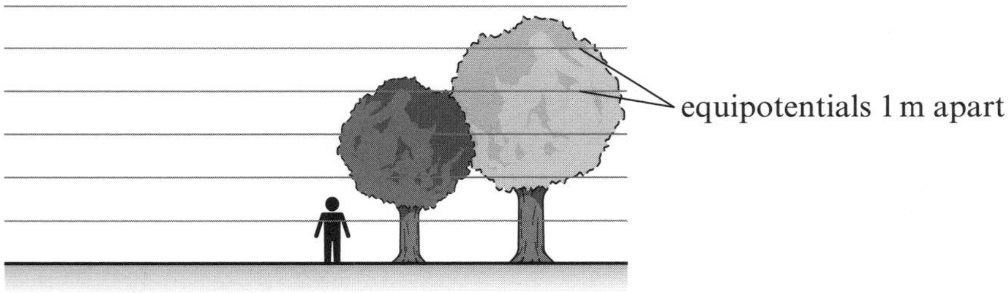

equipotentials 1 m apart

   (a) Label the equipotentials with values of gravitational potential difference from the surface (taken to be zero). The gravitational field strength near the surface is 9.81 N kg^{-1}.

   (b) Explain why, in this example, it is reasonable to assume that the gravitational field strength is constant.

....

.... **(2 marks)**

   (c) Add lines to the figure to show the gravitational field lines and equipotentials around the Earth.

(4 marks)

Energy changes in a gravitational field

1 (a) A cyclist is free-wheeling up a hill on a straight road that rises at an angle of 5.0° to the horizontal. His velocity at the bottom of the hill is $16\,\mathrm{m\,s^{-1}}$. The distance to the top of the hill is 125 m, measured along the road. The total mass of bicycle and rider is 90 kg. Can he reach the top of the hill without pedalling? Neglect the effects of friction.

> Start by working out the vertical height of the hill and use this to calculate the change in GPE from the bottom to the top.

....

....

....

.... **(4 marks)**

(b) A second cyclist approaches the hill at the same speed. She also free-wheels up the hill. However, the total mass of her cycle and herself is 85 kg. State and explain whether this will affect how far she can free-wheel up the slope. Once again, neglect the effects of friction.

....

> Think algebraically ... what cancels?

.... **(2 marks)**

2 Here is some data about geostationary satellites and the Earth.
- mass of Earth = $5.97 \times 10^{24}\,\mathrm{kg}$
- mass of satellite = 3000 kg
- radius of Earth = $6.37 \times 10^{6}\,\mathrm{m}$
- orbital period = 24 hours
- altitude of geostationary satellite (height above Earth's surface) = $3.58 \times 10^{7}\,\mathrm{m}$

(a) Use the data to calculate the gravitational potential at a point on the surface of the Earth and at the altitude of a geostationary satellite.

....

....

....

.... **(4 marks)**

(b) Use your answers to (a) to calculate the increase in gravitational potential energy of the satellite when it is lifted into orbit.

....

.... **(2 marks)**

⟩**Guided**⟩ (c) Calculate the gravitational potential energy of the satellite when it is in orbit.

E_g (on the ground) = mgh = ...

Total E_g (in orbit) = ... **(2 marks)**

Comparing electric and gravitational fields

1   Which of the following statements about electric and gravitational fields is incorrect?

   ☐ **A**   All masses accelerate at the same rate in the same uniform gravitational field.

   ☐ **B**   All charges accelerate at the same rate in the same uniform electric field.

   ☐ **C**   The electric field strength at a point is the vector sum of all electric fields superposed at that point.

   ☐ **D**   The gravitational field strength at a point is the vector sum of all gravitational fields superposed at that point.   **(1 mark)**

2   It is now known that in Rutherford's scattering experiment positively charged alpha particles scattered from positively charged gold nuclei because of the electrostatic repulsions between the two particles.

   Use the data below to show that Rutherford was justified in neglecting gravitational effects.

Mass of alpha particle:	6.64×10^{-27} kg
Mass of gold nucleus:	3.27×10^{-25} kg
Charge on alpha particle:	3.20×10^{-19} C
Charge on gold nucleus:	1.27×10^{-17} C

> If you approach this task algebraically you will be able to cancel out the particle separation.

....

....

....

....   **(4 marks)**

3   (a)   Draw one arrangement of charges that gives a positive electrostatic potential energy and explain why the electrostatic potential energy is positive.

....

....

....   **(3 marks)**

   (b)   Explain why it is not possible to create a positive gravitational potential energy.

....

....

....   **(2 marks)**

Orbits

1 (a) Show that the time period for a satellite orbiting a planet of mass M at distance R does not depend on the mass m of the satellite.

....

....

.... **(3 marks)**

(b) (i) Write down an expression for the gravitational potential energy of a satellite of mass m in a circular orbit of radius R around a planet of mass M.

.... **(1 mark)**

(ii) Derive an expression for the kinetic energy of a satellite of mass m in a circular orbit of radius R around a planet of mass M.

....

.... **(2 marks)**

(iii) Find the ratio of the gravitational potential energy to the kinetic energy of a satellite of mass m in a circular orbit around a planet of mass M.

.... **(1 mark)**

2 A spacecraft is in an elliptical orbit around the Earth similar to that shown in the figure.

(a) Describe the energy transfers that take place as the spaceship completes one orbit, starting and ending at point A.

....

....

.... **(3 marks)**

(b) Eventually, the spacecraft is going to leave the Earth's orbit and travel to a distant planet. Discuss whether it will require more fuel to do this if it leaves the orbit at A or at B.

| Consider the minimum |
| energy needed to |
| escape and the total |
| energy of the spacecraft |
| while in orbit. |

....

....

....

.... **(4 marks)**

Exam skills 12 Gravitational fields

1 (a) State Newton's law of gravitation.

.... **(1 mark)**

(b) Define gravitational field strength and state its S.I. units.

....

.... **(2 marks)**

(c) A satellite in a polar orbit has an orbital period of 2.0 hours. Calculate the altitude of the satellite above the surface of the Earth. (Mass of Earth = 6.0×10^{24} kg; radius of Earth = 6400 km.)

....

....

....

.... **(4 marks)**

(d) (i) What is meant by the gravitational potential at a point in space?

....

.... **(2 marks)**

(ii) Explain why all gravitational potential energies are negative.

....

....

.... **(3 marks)**

(e) The Moon's orbital radius is increasing at a rate of about 38 mm per year. Its present value is 3.9×10^8 m and the mass of the Moon is 7.3×10^{22} kg.

(i) Calculate the rate (in watts) at which the Moon's gravitational potential energy is increasing.

....

....

....

.... **(3 marks)**

(ii) Suggest how the total energy of the Earth–Moon system can still be conserved despite the increase in the Moon's gravitational potential energy.

....

.... **(2 marks)**

Simple harmonic motion

Practical skills

1   A student wants to measure the frequency of a simple pendulum. Describe a suitable method explaining how to ensure an accurate result.

....

....

....

....

.... **(5 marks)**

2   When a ruler is fixed to a bench so that part of it extends beyond the edge of the bench, the extended part will undergo vertical oscillations when displaced and released. A student is trying to find out whether the oscillations are simple harmonic. In order to do this she attaches a mass to the end of the ruler and measures its vertical displacement. She repeats the procedure for a range of masses.

She obtains the following data:

Mass/g	10	20	30	40	50	60	70
Deflection/mm	3.0	6.0	9.0	11.8	13.4	15.7	17.8

(a)   State the two conditions necessary for an oscillator to undergo simple harmonic oscillations.

....

.... **(2 marks)**

(b)   Discuss whether the ruler will undergo simple harmonic oscillations when displaced for the range of masses used. Consider the effect of small and large initial displacements of the end of the ruler.

....

....

....

.... **(4 marks)**

> Look at the data and see if it fits the criteria for S.H.M. in (a).

Guided

(c)   Suggest, giving reasons, how the time period of vertical oscillations would change if:

The ruler was moved so that a shorter length extended beyond the edge of the bench.

A shorter length would be stiffer, so ....

.... **(2 marks)**

Analysing simple harmonic motion

1   A simple harmonic oscillator is released from an initial displacement of 10 cm at time $t = 0$. It then oscillates with a time period of 0.20 s. Which of the equations below gives its displacement as a function of time?

☐ **A** $10 \cos (10\pi t)$

☐ **B** $20 \cos (10\pi t)$

☐ **C** $10 \cos (0.40\pi t)$

☐ **D** $20 \cos (0.40\pi t)$ **(1 mark)**

2   A simple harmonic oscillator of mass 0.25 kg has an amplitude of 8.0 cm and oscillates with a frequency of 2.0 Hz. At time $t = 0$ it has a positive displacement of 8.0 cm.

(a)   Calculate the time period of the oscillation.

....   **(1 mark)**

(b)   Calculate the displacement of the oscillator after:

(i)   0.125 s

....   **(1 mark)**

(ii) 0.25 s

....   **(1 mark)**

(iii) 0.40 s

....   **(1 mark)**

(c)   Calculate the maximum speed of the oscillator and state the value of displacement at which this occurs.

....

....   **(2 marks)**

(d)   Calculate the maximum force acting on the oscillator and state the displacements at which it occurs.

....

....   **(2 marks)**

⟩Guided⟩   (e)   Calculate the total energy of the oscillator.

Maximum KE (at maximum velocity) = ...

At this point, the potential energy is ....
                                                                (2 marks)

3   The displacement x of a simple harmonic oscillator is $x = A \cos (2\pi ft)$.
    Write down equations for the velocity and acceleration of this oscillator.

....

....   **(2 marks)**

Graphs of simple harmonic motion

1 The figure below shows how the displacement of a particular simple harmonic oscillator varies with time.

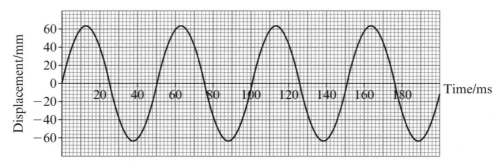

(a) State the amplitude of this oscillation.

.... **(1 mark)**

(b) Calculate the frequency of this oscillation.

.... **(1 mark)**

(c) Draw, using the axes below, a graph to show the variation of velocity with time for this oscillation. Add a suitable scale to the velocity axis.

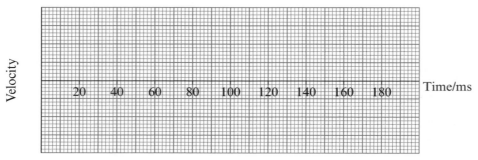

(4 marks)

(d) Draw, using the axes below, a graph to show the variation of acceleration with time for this oscillation. Add a suitable scale to the acceleration axis.

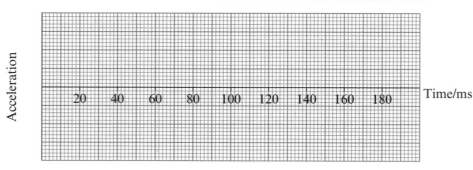

(4 marks)

The mass–spring oscillator and the simple pendulum

1 Which line in the table correctly indicates the changes, if any, to the time periods of a mass–spring oscillator and a simple pendulum if they are both taken to the Moon?

	Period of mass–spring system	Period of simple pendulum
A	increases	stays the same
B	decreases	stays the same
C	stays the same	increases
D	stays the same	decreases

(1 mark)

2 A student sets up an experiment to measure the time period of a mass–spring oscillator. He uses a steel spring of spring constant $25\,\text{N m}^{-1}$ attached to a mass of $400\,\text{g}$.

(a) Calculate the expected time period for this oscillator.

....

.... **(2 marks)**

> **Guided**

(b) When he analyses his results, he finds that the time period is significantly greater than the expected theoretical value. This is not a result of experimental error. Suggest and explain a physical reason for this difference.

The spring itself has mass, ...

....

.... **(3 marks)**

3 Some old clocks use a simple pendulum to keep time. Whilst such clocks are often beautiful and can be quite accurate, there are also problems linked to the behaviour of the simple pendulum itself.

(a) Calculate the length of a simple pendulum that would have a time period of exactly $1.00\,\text{s}$.

....

.... **(2 marks)**

(b) Temperature changes can affect the length of a pendulum. Calculate the percentage change in time period when the length of the pendulum in (a) increases by 0.50% as a result of thermal expansion.

....

....

....

> You could use the rule for combining fractional uncertainties: in this case $\dfrac{\Delta T}{T} = \dfrac{1}{2}\dfrac{\Delta l}{l}$.

(3 marks)

Energy and damping in simple harmonic oscillators

1   The frequency and amplitude of vibration of a loudspeaker are both doubled.
    By what factor does the energy of the oscillations change?

   ☐ **A** $\times 2$ ☐ **B** $\times 4$

   ☐ **C** $\times 8$ ☐ **D** $\times 16$ **(1 mark)**

2   A mass of 500 g is suspended from a spring of spring constant $42\,N\,m^{-1}$. It is lifted
    2.0 cm vertically and then released; it oscillates with an initial amplitude of 2.0 cm.

   (a) What energy transfers take place during one complete oscillation?

   ....

   .... **(2 marks)**

   (b) Calculate the extension of the spring when the mass is at its equilibrium
        position.

   .... **(2 marks)**

   (c) Calculate the energy stored in the spring when the
        mass is at:

   > The energy stored in stretched spring is given by $E = \frac{1}{2}kx^2$.

   (i) its equilibrium position

   .... **(1 mark)**

   (ii) 2.0 cm above its equilibrium position

   .... **(1 mark)**

   (iii) 2.0 cm below its equilibrium position

   .... **(1 mark)**

   (d) Calculate the change in gravitational potential energy of the mass between
        the top and bottom of its first oscillation and hence show that the loss of
        gravitational energy is equal to the gain in elastic potential energy between these
        two positions.

   ....

   .... **(2 marks)**

⟩**Guided**⟩   (e) Calculate the maximum kinetic energy of the oscillator
             and hence calculate its maximum velocity.

   > Maximum KE is equivalent to maximum GPE.

   maximum KE = ...

   $KE = \frac{1}{2}mv^2$ so $v =$ ....

   .... **(3 marks)**

Forced oscillations and resonance

1 A student carries out an experiment to investigate the response of a mass–spring oscillator to a driving force using the apparatus shown in the figure below.

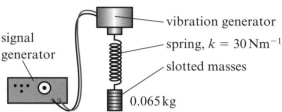

signal generator

vibration generator

spring, $k = 30\,Nm^{-1}$

slotted masses

0.065 kg

(a) (i) State what is meant by **natural frequency** and **forcing frequency**.

....

....

....

.... **(2 marks)**

(ii) Calculate the natural frequency of oscillations of this mass–spring oscillator.

....

.... **(1 mark)**

(b) (i) In the experiment, the student sets the signal generator to a small fixed amplitude A_0 and then gradually increases the frequency from 0 to 10 Hz. Use the axes below to show the expected response of the oscillator. Assume that damping forces are small (but not zero).

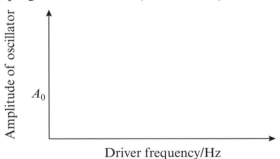

Amplitude of oscillator

A_0

Driver frequency/Hz

(3 marks)

(ii) Explain what is meant by **resonance** and state when it occurs.

....

.... **(2 marks)**

⟩**Guided**⟩ (iii) Describe the energy transfers taking place at resonance and their effects.

At resonance, the driver continually transfers energy to the driven oscillator,
causing its amplitude to grow. At the same time, work done by the oscillator

....

....

....

.... **(4 marks)**

133

Driven oscillators

1 A student carries out an experiment to investigate the response of a mass–spring oscillator to a driving force using the apparatus shown in the figure below.

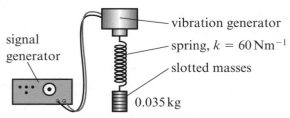

signal generator

vibration generator

spring, $k = 60\,\mathrm{N m^{-1}}$

slotted masses

0.035 kg

(a) The student sets the signal generator to a small fixed amplitude A_0 and then gradually increases the frequency from 0 to 10 Hz. Sketch the expected response of the oscillator on the axis below. Mark a scale on the frequency axis. Assume that damping forces are negligible.

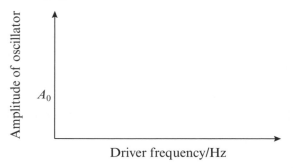

Amplitude of oscillator

A_0

Driver frequency/Hz

(4 marks)

(b) In a second experiment, the student sticks a horizontal cardboard disc to the bottom of the masses to act as a damper. Draw a second line on the graph above to show the effect of damping.

(2 marks)

(c) The student wishes to change the mass–spring system so that resonance does not occur in the range 0–10 Hz. Suggest and explain two ways in which this could be done.

> Think about the equation for the natural frequency of the oscillator. What changes would increase this frequency?

....

....

....

.... **(3 marks)**

Exam skills 13
Simple harmonic motion

1 Modern mountain bikes have suspension springs on their front forks, so that the front wheel can move up and down relative to the frame. The spring on a particular mountain bike has a spring constant of $6500 \, \text{N} \, \text{m}^{-1}$ and it compresses by $0.084 \, \text{m}$ when the bike hits a bump. The mass of bike and rider is $75 \, \text{kg}$.

(a) Assuming that the bike and rider system together act like a mass–spring oscillator, calculate:

(i) the natural frequency of simple harmonic oscillations after hitting the bump.

....

....

.... **(3 marks)**

(ii) the maximum energy stored in the spring.

....

.... **(2 marks)**

(iii) the maximum vertical acceleration of the bike during the oscillations.

....

.... **(2 marks)**

(b) the rider has a mass of $60 \, \text{kg}$. Calculate the maximum vertical reaction force from his saddle after he hits the bump.

....

.... **(2 marks)**

(c) (i) Explain why a damper is also included in the front forks.

....

.... **(2 marks)**

(ii) Sketch, on the axes below, a graph to indicate how the compression of the spring varies from the moment the bike hits the bump until the oscillations have been completely damped after about 2 complete cycles of oscillation.

(4 marks)

ANSWERS

1. S.I. units

1 C − charge is not one of the S.I. base units kg, m, s, K, A, mol and cd (candela, the unit of luminous intensity, which you do not need to learn about). **(1)**

2 D − work is force × distance and force is mass × acceleration, therefore units are $\mathrm{kg\,m\,s^{-2}} \times \mathrm{m} = \mathrm{kg\,m^2\,s^{-2}}$. **(1)**

3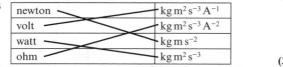
(3)

4 Force × time has units $\mathrm{kg\,m\,s^{-2}} \times \mathrm{s} = \mathrm{kg\,m\,s^{-1}}$ and mass × velocity has units $\mathrm{kg} \times \mathrm{m\,s^{-1}}$, which is also $\mathrm{kg\,m\,s^{-1}}$, so they are both the same. **(2)**

5 R is just a distance with units of m. $\frac{v^2}{g}$ has units $(\mathrm{m\,s^{-1}})^2/\mathrm{m\,s^{-2}} = \mathrm{m^2\,s^{-2}}/\mathrm{m\,s^{-2}}$, which is also m. $\sin(2\theta)$ does not have any units, so both sides have the same units. **(2)**

6 (a) Rearranging the equation gives $\rho = \frac{RA}{l}$, and considering the units of the quantities on the right-hand side gives units $\Omega\,\mathrm{m^2}/\mathrm{m} = \Omega\,\mathrm{m}$. **(2)**

   (b) $R = \frac{V}{I} = \frac{P}{I^2}$, so units are $\mathrm{W\,A^{-2}} = \mathrm{kg\,m^2\,s^{-3}\,A^{-2}}$. **(2)**

2. Practical skills

1 D − $5.12 \pm 0.02\,\mathrm{mm}$ **(1)**

2 B − $0.13 \pm 0.002\,\mathrm{mm}$ **(1)**

3 (a) An error is any deviation of a measurement from its true value. **(1)**

   (b) A random error causes measurement results to vary in an unpredictable way whereas a systematic error causes measurement results differing from the true value by a consistent amount each time. **(2)**

   (c) A random error has an equal chance of being positive or negative, so when a mean is taken they will tend to cancel out. A systematic error is always the same, so it would still be present when an average has been taken. **(2)**

   (d) (i) The presence of an unexplained non-zero intercept might suggest a systematic error. **(1)**

   (ii) The points would be significantly scattered above and below the line of best fit. **(1)**

3. Estimation

Your numerical estimates may vary, but your reasoning should be along these lines:

1 Let's assume the town has 20 000 residents in 5000 homes plus shops, factories and schools, etc. There will be periods of high demand, e.g. evenings, and low demand, e.g. at night. In the day, people may be away from home but will still account for a share of what is being used in the town. A reasonable estimate per person would be about a kilowatt averaged over the day so about $20\,000 \times 1000 = 2 \times 10^7\,\mathrm{W}$ or 20 MW. **(3)**

2 An object falling 2 m will acquire a velocity of $v = \sqrt{2 \times 9.81 \times 2} = 6.3\,\mathrm{m\,s^{-1}}$ (using $v^2 - u^2 = 2as$). If you bend your legs as you land and stop over ~0.5 m, your deceleration is $\frac{v^2}{2s} = 39.2\,\mathrm{m\,s^{-2}}$. If your mass is 80 kg, using $F = ma$ gives a decelerating force of 3100 N and in addition your weight is another 780 N, so a total force of a little under 4000 N. **(3)**

3 A train of 12 carriages, each with the mass of about 10 cars, has mass ~200 000 kg (200 tonnes). An express train may travel at $180\,\mathrm{km\,s^{-1}}$, or $50\,\mathrm{m\,s^{-1}}$. Its KE is $\frac{1}{2}mv^2 = 0.5 \times 200\,000 \times 50^2 = 250\,\mathrm{MJ}$, so yes, the beam of the LHC has a comparable amount of energy. **(2)**

4 You can fill a kettle that holds a litre ($1000\,\mathrm{cm^3}$) in about 3 s. A swimming pool $25\,\mathrm{m} \times 10\,\mathrm{m} \times 2\,\mathrm{m}$ has a volume of $500\,\mathrm{m^3}$. There are $10^6\,\mathrm{cm^3}$ in $1\,\mathrm{m^3}$, so there are 1000 litres in $1\,\mathrm{m^3}$ and it would take $500 \times 1000 \times 3 = 1.5$ million seconds to fill the pool, or about 17 days. **(3)**

4. SUVAT equations

1 (a) If the time is t, B travels $25t$ metres. A (using $s = ut + \frac{1}{2}at^2$) travels $(25t + \frac{1}{2} \times 2 \times t^2)$ metres. A must travel $10 + 15 + 5 + 5 = 35\,\mathrm{m}$ farther than B.
$35 + 25t = (25t + \frac{1}{2} \times 2 \times t^2)$
$t^2 = 35$ and so $t = 5.92\,\mathrm{s}$ or just under 6 s.

   (b) B travels $25 \times 5.92 = 148\,\mathrm{m}$ (or use $t = 6\,\mathrm{s}$ to get 150 m) **(1)**.

   (c) A travels $148 + 35 = 183\,\mathrm{m}$ (or $150 + 35 = 185\,\mathrm{m}$). **(1)**

   (d) $v = u + at = 25 + 2 \times 5.92 = 36.8\,\mathrm{m\,s^{-1}}$ (or $v = 37\,\mathrm{m\,s^{-1}}$) **(2)**

2 (a) At maximum height, the ball will have zero velocity v, so using $v^2 - u^2 = 2as$ gives $-30^2 = 2 \times (-9.81) \times s$, so $s = 46\,\mathrm{m}$. **(2)**

   (b) When the ball returns to its starting height, $s = 0$ and $0 = 30t + \frac{1}{2} \times (-9.81) \times t^2$, so $t = 0$ (i.e. the launch time) or $t = 6.1\,\mathrm{s}$. **(2)**

5. Displacement–time, velocity–time and acceleration–time graphs

1 (a)
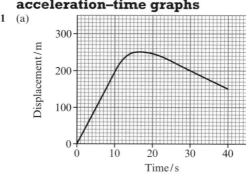
(4)

   (b)
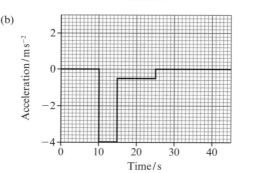
(4)

   (c) distance (scalar) $= 200 + 50 + 25 + 75 = 350\,\mathrm{m}$ **(2)**

6. Scalars and vectors

1

Quantity	Vector	Scalar
distance		✓
momentum	✓	
speed		✓
energy		✓

(4)

2 (a) Scalar quantities are just magnitudes, so adding them up can only produce one answer, but vectors also have direction. The result of adding two vectors thus depends on their relative directions as well as their magnitudes. **(2)**

   (b) 0 N ⟨10 N⟩ ⟨30 N⟩ ⟨70 N⟩ 80 N **(2)**

   (c) A − Displacement and velocity are vector quantities, but time is a scalar quantity. **(1)**

   (d) D − Her total displacement for the two journeys was zero (she has returned to the place she started from), so her average velocity was $0\,\mathrm{km\,h^{-1}}$ over the two journeys. **(1)**

3 Bikram is right, because temperature going up or down is just a change in magnitude; there is no physical direction involved. **(2)**

7. Resolution of vectors

1 (a)

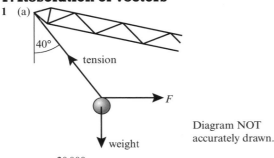

Diagram NOT accurately drawn. **(2)**

(b) $T = \dfrac{20\,000}{\cos 40°} = 26\,100\text{ N (26.1 kN)}$ **(3)**

(c) $F = 26\,100 \times \cos 50° = 16\,800\text{ N (16.8 kN)}$ **(2)**

(d) acceleration $a = \dfrac{W \cos 50°}{m_{\text{ball}}} = g \times \cos 50° = 6.3\text{ m s}^{-2}$ **(2)**

2 B: $F + W \sin \theta$ **(1)**

8. Adding vectors

1 (a) The duck will take 10 s to paddle 2.0 m relative to the water. In this time, it will have been carried 1.0 m downstream, so its displacement will be the vector sum of **AB** and **BC**. The resultant displacement has a magnitude of $\sqrt{1.0^7 + 2.0^2} = 2.24$ m. The direction is at an angle of $\tan^{-1}\left(\dfrac{1.0}{2.0}\right) = 26.6°$. **(2)**

(b) The direction should be at an angle of $\sin^{-1}\left(\dfrac{0.10}{0.20}\right) = 30°$ to **AB**. **(2)**

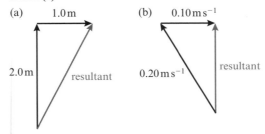

(c) The duck has to paddle $\dfrac{2}{\cos 30°} = 2.31$ m. The time taken will be $\dfrac{2.31}{0.20} = 11.5$ s **(2)**

2 Draw a triangle (or a parallelogram) to scale:

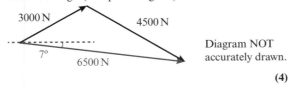

Diagram NOT accurately drawn.

(4)

9. Projectiles

1 (a) Using $s = ut + \frac{1}{2}at^2$ leads to $t = \sqrt{\dfrac{2s}{a}} = \sqrt{\dfrac{2 \times 80}{9.81}} = 4.04$ s. **(2)**

(b) Range $= 480 \times 4.04 = 1940 \approx 1900$ m **(2)**

(c)

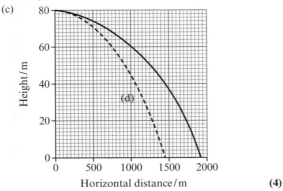

(4)

(d) As above − reduced horizontal range **(2)**

2 r will be a maximum when $\sin 2\theta$ is a maximum. The sine function has a maximum value of 1 when the angle is 90°. When $\sin 2\theta = 1$, it follows that $2\theta = 90°$ and $\theta = 45°$ for maximum range. **(2)**

10. Free-body force diagrams

1 (a)

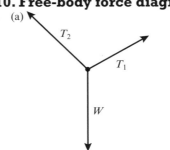

(3)

(b) As the mirror is in equilibrium (both horizontally and vertically), the weight downwards must be balanced by a similar upward force, i.e. the sum of the two forces due to the tensions in the string. **(2)**

(c)

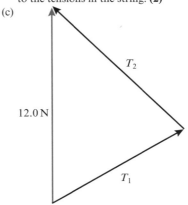

$T_1 = 8.8$ N and $T_2 = 10.7$ N **(4)**

11. Newton's first and second laws of motion

1 (a) According to Newton's first law of motion, in the absence of a force acting on it, an object will move in a straight line at constant speed. The converse also applies, so an object moving in a straight line at constant speed must have zero resultant force acting on it. **(2)**

(b) As the car is not moving in a straight line, it must be subject to a resultant force, which in this case causes a change in direction rather than a change in speed. **(2)**

(c) D **(1)**

2 (a)

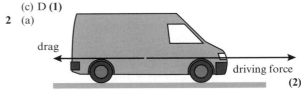

(2)

(b) Using $\Sigma F = ma$ and rearranging to give $a = \dfrac{\Sigma F}{m}$, we have $a = \dfrac{4400 - 800}{4000} = 0.9\text{ m s}^{-2}$. **(2)**

(c) The maximum speed is achieved when there is no resultant force and the driving force = drag = 4400 N. The maximum speed will be $20 \times \sqrt{\dfrac{4400}{800}} = 47\text{ m s}^{-1}$ **(2)**

12. Measuring the acceleration of free fall

1 (a) For uniformly accelerated motion, we can use $s = ut + \frac{1}{2}at^2$. In this case, $h = \frac{1}{2}at^2$. If we compare this with the equation of a straight line, $y = mx + c$, we can see that if h is the y-variable and t^2 is the x-variable, we will get a straight line. **(2)**

(b) The gradient of the graph is equal to $\frac{1}{2}a$, so if we find the gradient of the graph and double it, we can find the acceleration of the ball. **(2)**

(c)

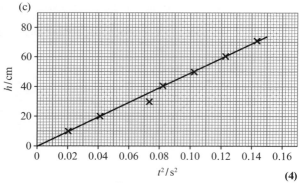

h/cm axis, *t²*/s² axis

(4)

(d) The gradient $= \dfrac{70}{0.143} = 490$, so the acceleration is $490 \times 2 = 980\,\text{cm s}^{-2}$ or $9.8\,\text{m s}^{-2}$. **(2)**

13. Newton's third law of motion

1 (a)

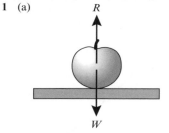

R (up), *W* (down)

(2)

(b) The weight of the apple is the gravitational pull of the Earth on the apple, so according to Newton's third law, there will be an equal and opposite pull of gravity on the Earth due to the apple. **(1)**

2 The helicopter's blades push down on the air. Newton's third law of motion says that for every force there is an equal and opposite force, so the helicopter can hover because the downward force acting on the air due to the helicopter blades is equal to the upward force of the air acting on the helicopter blades. This upward force is equal to the downward force of gravity acting on the helicopter. **(2)**

3 (a) Using $\Sigma F = ma$ gives $F = 1400 \times 2.5 = 3500\,\text{N}$ to the right. **(2)**

   (b) 3500 N to the left **(2)**

   (c) They are both frictional forces. **(2)**

14. Momentum

1 B – Check that you have converted all values to base units. **(1)**

2 C – P is stationary and Q moves off at $2\,\text{m s}^{-1}$. **(1)**

3 D – $0.5\,\text{m s}^{-1}$ to the left, so total momentum is still $0\,\text{kg m s}^{-1}$. **(1)**

4 Conservation of momentum gives total momentum before impact = total momentum after impact = $0.200 \times 31 = 0.200 \times v_{\text{club}} + 0.045 \times 60$, so $v_{\text{club}} = 17.5\,\text{m s}^{-1}$ to the right. **(2)**

15. Moment of a force

1 The centre of mass must be vertically below the point of suspension, otherwise a resultant moment will act until the previous condition is satisfied.

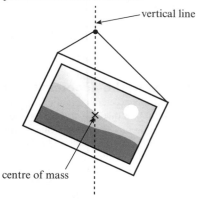

vertical line

centre of mass

(3)

2 (a) For an object to be in rotational equilibrium, the total clockwise moment must be equal to the total anticlockwise moment. **(2)**

(b) Considering moments about the wheel of the sack truck, $2.0 \times 0.6 = F \times 1.5$, so $F = \dfrac{2.0 \times 0.6}{1.5} = 0.8\,\text{kN}$ (800 N). **(2)**

(c) Either principle of moments pivoting about the centre of mass of the sack truck: $F \times 0.6 = 0.8 \times 0.9$ giving $F = 1.2\,\text{kN}$ or total downward forces = total upward forces; $F = 2.0 - 0.8 = 1.2\,\text{kN}$ acting vertically downwards. **(2)**

16. Exam skills 1 Forces, moments and equilibrium

1 (a) Taking moments about the pivot, the force acting at the support must balance the weight of the barrier acting at its centre of mass: $F_{\text{support}} \times 4.0 = 200 \times 2.0$; $F_{\text{support}} = 100\,\text{N}$. **(2)**

(b) The barrier is in force equilibrium, so the downward weight of the barrier must be balanced by the upward force due to the support and the pivot. **(1)**

(c) When the barrier starts to lift, there is zero force between the support and the barrier. If we take moments about the pivot, we get $2.0 \times T\cos 45° = 2.0 \times 200$ and $T = \dfrac{200}{\cos 45°} = 280\,\text{N}$. **(3)**

(d) The force at the pivot must balance the horizontal component of the force due to the tension in the wire rope. **(1)**

(e) As the barrier approaches the vertical, the angle between the wire rope and the barrier approaches 90°. This means the component of the tension at right angles to the barrier increases while at same time, the component of the weight of the barrier at right angles to the barrier decreases. Both effects reduce the tension in the wire rope. **(3)**

17. Work

1 (a) 400 N **(1)**

(b)

friction, 400 N

(1)

(c) $W = Fs = 400 \times 5.0 = 2000\,\text{J}$ **(3)**

(d) 2000 J **(1)**

(e) Because the pulling force is not horizontal, only the horizontal component, $460\cos 30°$, of the tension in the rope acts against friction, so the total tension must be greater. **(2)**

(f) $W = Fs = (460\cos 30°) \times 5.0 = 1992 \approx 2000\,\text{J}$ **(2)**

(g) They are the same because the work done only depends on the frictional force and the distance moved, and both of these are the same in both cases. **(2)**

18. Kinetic energy and gravitational potential energy

1 (a) $E_k = \tfrac{1}{2}mv^2 = 0.5 \times 1800 \times 20^2 = 360\,000\,\text{J}$ **(1)**

(b) The work done by the brakes is equal to the initial kinetic energy of the car = Fs; this leads to $F = \dfrac{360\,000}{40} = 9000\,\text{N}$. **(2)**

(c) The kinetic energy of the car is transferred by heating to the brakes. **(2)**

2 (a) After each bounce, the maximum velocity is reduced, so the ball has less kinetic energy. **(2)**

(b) If the ball has less KE when leaving contact with the surface, it will have less PE at the top of the bounce, and as $E_p = mgh$, the height h must be less. **(2)**

(c) Height of first bounce is 0.60 m which is 60% of the 1 m from which it was dropped, so PE has decreased by 40% and so KE has also decreased by 40%.
Alternatively, velocity decreases from $4.5\,\text{m s}^{-1}$ to $3.5\,\text{m s}^{-1}$ on bouncing, therefore KE decreases to $\left(\dfrac{3.5}{4.5}\right)^2 = 0.60$ or 60% of initial value. It must therefore have lost 40% of its initial KE. **(2)**

19. Conservation of energy

1 (a) Energy cannot be created or destroyed. It can only be transferred from one form to another. (Alternatively: The total energy of a closed system remains constant.) **(2)**
(b) When energy is transferred, some of it is usually transferred into heat energy. This energy is lost to the surroundings and cannot be reused. This fact means that our supply of useful energy is finite and should not be wasted. **(2)**

2 (a) $W = Fs = 200 \times 0.45 = 90\,\text{J}$ **(1)**
(b) 60% of 90 J is $0.60 \times 90 = 54\,\text{J}$ **(1)**
(c) The remaining stored energy is transferred to sound, KE of the bow, the archer and heat. **(2)**
(d) $E_k = \frac{1}{2}mv^2$. Rearranging gives $v = \sqrt{\dfrac{2E_k}{m}} = \sqrt{\dfrac{2 \times 54}{0.027}}$ $= 63\,\text{m s}^{-1}$ **(2)**

20. Work and power

1 (a) Work done = gain in GPE of weight = $mg\Delta h$ $= 168 \times 9.81 \times 1.50 = 2470\,\text{J}$ **(1)**
(b) $P = \dfrac{W}{t} = \dfrac{2470}{5} = 494\,\text{W}$ **(2)**
(c) During the parts of the lift when the weight was moving more quickly, the rate of doing work, i.e. the power, would have been much greater. When the weight was stationary, no work was being done, so the average power was lower than the peak power. **(2)**

2 The rate of doing work is the rate of gain of GPE of the lift, i.e. $P = \dfrac{mg\Delta h}{\Delta t}$ but $\dfrac{\Delta h}{\Delta t}$ is the vertical speed so $\dfrac{\Delta h}{\Delta t} = \dfrac{P}{mg} = \dfrac{90\,000}{1800 \times 9.81} = 5.1\,\text{m s}^{-1}$. **(2)**

3 (a) $E_k = \frac{1}{2}mv^2 = 0.5 \times 0.0080 \times 700^2 = 1960\,\text{J}$ **(1)**
(b) The above energy is attained in the time it takes the bullet to travel down the barrel of length 0.52 m at an average speed of $350\,\text{m s}^{-1}$ ($[v - u]^2$) so $t = \dfrac{0.52}{350} = 0.00149\,\text{s}$. This requires a power of $P = \dfrac{E}{t} = \dfrac{1960}{0.00149} = 1.3 \times 10^6\,\text{W}$. **(2)**

21. Efficiency

1 (a) The time taken to travel 40 km at $20\,\text{m s}^{-1} = \dfrac{40\,000}{20} = 2000\,\text{s}$.
The output energy is $12.0\,\text{kW} \times 2000\,\text{s} = 24.0\,\text{MJ}$. **(2)**
(b) If the efficiency is 20%, the input energy is $\dfrac{24.0}{0.20} = 120\,\text{MJ}$.
1.0 kg of fuel provides 40 MJ, so $\dfrac{120}{40} = 3.0\,\text{kg}$ of fuel is needed. **(2)**

2 (a) $100 - 73 - 8 - 7 = 12\%$ **(1)**
(b) Electrical output = 12% of $750\,\text{W m}^{-2} = 0.12 \times 750$ $= 90\,\text{W m}^{-2}$. The area required for 1.8 kW is $\dfrac{1800}{90} = 20\,\text{m}^2$. **(2)**

22. Exam skills 2 Forces, energy and motion

1 (a) Increase in GPE= $mgh = 2500 \times 9.81 \times 4.0 = 9.8 \times 10^4\,\text{J}$ **(2)**
(b) $9.8 \times 10^4\,\text{J}$ **(1)**
(c) Work done by force between pile and ground = $F \times 0.12$ $= 0.80 \times 9.8 \times 10^4$ so $F = 6.5 \times 10^5\,\text{N}$. **(3)**
(d) thermal energy **(1)**
(e) Average power output $= \dfrac{40}{60} \times 120 \times 10^3 = 8.0 \times 10^4\,\text{W}$. **(2)**

23. Basic electrical quantities

1 (a) The current flowing through part of a circuit is equal to the rate of flow of electrical charge through it. **(1)**
(b) The potential difference between two points in a circuit is the energy transferred per unit charge passing between those points. **(1)**

2 (a) $Q = It = 0.050 \times 100 = 5.0\,\text{C}$ **(2)**
(b) $W = VIt = VQ = 9 \times 5.0 = 45\,\text{J}$ **(2)**
(c) $E = VQ = 6 \times 5.0 = 30\,\text{J}$ **(1)**
(d) Energy must be conserved, so if the battery is producing 9 J of electrical energy per coulomb of charge, and 6 J is converted to heat and light by the lamp per coulomb, there must be $9 - 6 = 3\,\text{J}$ remaining per coulomb and this must be equal to the potential difference across the resistor. **(3)**
(e) $R = \dfrac{V}{I} = \dfrac{6}{0.050} = 120\,\Omega$ **(1)**

24. Ohm's law

1 Resistance is defined as the ratio of the potential difference across a conductor to the current through it $\left(\text{or } R = \dfrac{V}{I}\right)$. **(1)**

2 (a) Variable resistor — used to control the current through the tank and hence the potential difference between the electrodes. **(2)**
(b) The depth to which the copper electrodes are immersed; the concentration of the copper sulfate solution; the separation of the copper electrodes. **(2)**
(c)

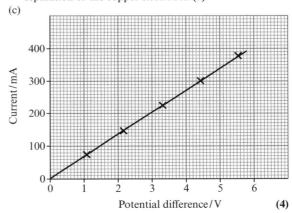

(4)
(d) The graph is a straight line through the origin, so it is correct to say that the current is directly proportional to potential difference, that is, it obeys Ohm's law. **(2)**

25. Conservations laws in electrical circuits

1 (a)

LED combination	I_1 / mA	I_2 / mA	I_3 / mA	I_4 / mA
red	20	20	0	0
red + amber	45	20	25	0
green	18	0	18	18
amber	25	0	25	0

(4)
(b) The p.d. across R is the e.m.f. of the battery minus the total p.d. across the three LEDs in series. The p.d. across R is $9.0 - (3 \times 2.2) = 2.4\,\text{V}$. **(1)**
(c) The current through R is 20 mA = 0.020 A. $R = \dfrac{V}{I} = \dfrac{2.4}{0.020} = 120\,\Omega$. **(2)**

26. Resistors

1 If each resistor has resistance R, the total resistances are R, $\dfrac{2R}{3}$, $\dfrac{3R}{3}$ and $\dfrac{6R}{3}$, respectively, so:
(a) C has the highest resistance. **(1)**
(b) B has the lowest resistance. **(1)**

2 (a) 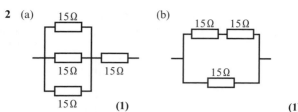 (b) **(1)** **(1)**

3 (a) The network is equivalent to $10 + 15 + 20 = 45\,\Omega$ in parallel with $25 + 30 = 55\,\Omega$ giving a total resistance of $\frac{45 \times 55}{45 + 55} = 24.75\,\Omega$ or $25\,\Omega$ to 2 significant figures. **(1)**

(b) The network is equivalent to $10 + 20$, 30 and $5 + 25$ in parallel and as this is three lots of $30\,\Omega$ in parallel, the result is $\frac{30}{3} = 10\,\Omega$. **(1)**

27. Resistivity

1 (a) Using $R = \frac{V}{I}$ gives $R = \frac{230}{4.35} = 52.9\,\Omega$. **(1)**

(b) Using $A = \frac{\pi D^2}{4}$ gives $A = \frac{\pi \times (0.31 \times 10^{-3})^2}{4}$
$= 7.55 \times 10^{-8}\,\mathrm{m}^2$ **(1)**

(c) $R = \frac{\rho l}{A}$ can be rearranged to give $l = \frac{RA}{\rho}$
$= \frac{52.9 \times 7.55 \times 10^{-8}}{1.06 \times 10^{-6}} = 3.77\,\mathrm{m}$. **(3)**

2 (a) $5.0\,\mu\mathrm{m} = 5.0 \times 10^{-6}\,\mathrm{m}$ and the twelve wires are in series, so using $R = \frac{\rho l}{A}$ gives $\frac{4.9 \times 10^{-7} \times 0.010 \times 12}{0.10 \times 10^{-3} \times 5.0 \times 10^{-6}} = 120\,\Omega$. **(3)**

(b) The volume is constant, so $V = Al = A \times 0.010$
$= A' \times 0.01001$, where A' is the area of cross-section after straining. $A' = A \times \frac{0.010}{0.01001} = 0.999\,A$, so A has decreased by 0.1%. **(2)**

28. Resistivity measurement

1 (a)

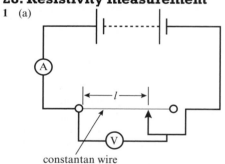

constantan wire **(2)**

(b) **(4)**, (c)

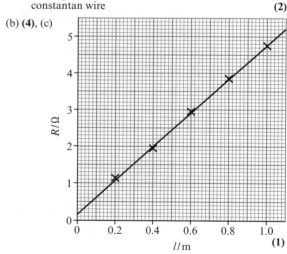

(1)

(d) Best-fit line (not through origin!) has gradient $\frac{4.6}{1.0} = 4.6\,\Omega\,\mathrm{m}^{-1}$. **(2)**

(e) Diameter is 0.37 mm, so area is $A = \frac{\pi D^2}{4}$
$= \frac{\pi \times (0.37 \times 10^{-3})^2}{4} = 1.08 \times 10^{-7}\,\mathrm{m}^2$. **(1)**

(f) As $\rho = \frac{RA}{l}$ and the gradient is $\frac{R}{l}$, it follows that $\rho = $ gradient $\times A = 4.6 \times 1.08 \times 10^{-7} = 5.0 \times 10^{-7}\,\Omega\,\mathrm{m}$. **(2)**

29. Current equation

1 (a) $I = nqvA$, so $v = \frac{I}{nqA}$
$= \frac{10}{8.4 \times 10^{28} \times 1.60 \times 10^{-19} \times 1.0 \times 10^{-6}}$
$= 7.4 \times 10^{-4}\,\mathrm{m\,s}^{-1}$. **(2)**

(b) A copper wire contains free electrons throughout its entire volume. Turning on a switch and completing a circuit affects them all at the same time, so those nearest the lamp will cause the lamp to light straight away. There is no need to wait for electrons to move any great distance around the circuit. **(2)**

(c) The density of charge carriers in silicon is several orders of magnitude less than that of electrons in copper, so the number of available charge carriers that can contribute to the flow of current in copper is vastly greater than the number in silicon. This means that for a given applied potential difference, the current in silicon will be much less and hence the resistivity will be much greater. **(3)**

2 The number of charge carriers in a metal such as a lamp filament is largely independent of temperature; however, the overriding phenomenon that determines the resistivity of a metal is how the conduction electrons interact with the metal lattice. As temperature increases, so does lattice vibration, leading to greater resistivity. In a semiconductor, such as in a negative temperature coefficient thermistor, the number of charge carriers increases dramatically with temperature, resulting in a significant decrease in resistivity. **(3)**

30. E.m.f. and internal resistance

1 (a) The e.m.f. of a battery is equal to the energy transferred from chemical energy to electrical energy per unit of charge passing through it, $6.0\,\mathrm{J\,C}^{-1}$ in this instance. **(2)**

(b) The internal resistance of a battery is the resistance of a power source. It behaves as if it were in series with the e.m.f. of the battery, in this instance like a $0.2\,\Omega$ resistor. **(1)**

(c) When there is internal resistance (and a current flowing), some p.d. is always 'dropped' across it, which means that the terminal voltage will be less than 6.0 V and the maximum power and hence maximum brightness associated with the normal rating of the bulb cannot be achieved. **(2)**

2 (a)

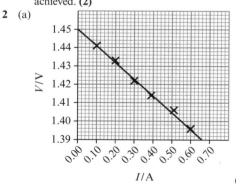

(5)

(b) In this experiment, the gradient of the graph is equal to $-r$ so, as the gradient is negative, r is a positive quantity.
The gradient is equal to $\frac{-0.06}{0.60} = -0.09\,\Omega$, so the internal resistance is $0.09\,\Omega$. **(2)**

(c) The e.m.f. of the cell is equal to the y intercept or 1.45 V. **(1)**

31. Potential divider circuits

1 (a) The potential divider formula $V_1 = \dfrac{VR_1}{R_1 + R_2}$ gives
$V = \dfrac{9 \times 30}{20 + 3} = 5.4\,\text{V}$ **(2)**

(b) The p.d. across the $20\,\Omega$ resistor below Y is $9.0 - 5.4 = 3.6\,\text{V}$, so the p.d. between X and Y is $5.4 - 3.6 = 1.8\,\text{V}$. **(2)**

2 (a) The p.d. across the $10\,\text{k}\Omega$ resistor is $\dfrac{6 \times 10}{10 + 2} = 2.0\,\text{V}$. **(1)**

(b) The voltmeter is in parallel with the $10\,\text{k}\Omega$ resistor, so the effect is to produce a potential divider consisting of a $20\,\text{k}\Omega$ resistor in series with a $5\,\text{k}\Omega$ resistor and the voltmeter will read $\dfrac{6 \times 5}{5 + 20} = 1.2\,\text{V}$. **(2)**

32. Exam skills 3 Circuit analysis

1 (a) Electromotive force is the energy transferred by a battery into electrical form as electrical charge passes through it. **(2)**

(b) The maximum current is $I = \dfrac{\varepsilon}{r} = \dfrac{12.0}{1.0} = 12\,\text{A}$. **(1)**

(c) The terminal p.d. is $10.0\,\text{V}$. **(1)**

(d) When current flows, there will be a potential difference across the internal resistance or 'lost volts', which must be subtracted from the e.m.f. in order to determine the potential difference across the terminals of the battery. In this case, $2.0\,\text{V}$ is dropped across the internal resistance of the battery. **(2)**

(e) The p.d. across the circuit is $10\,\text{V}$. The current is equal to the lost volts divided by the internal resistance $= \dfrac{2.0}{1.0} = 2.0\,\text{A}$. **(2)**

(f) When the second lamp is added in parallel to the first, more current flows, so the size of the lost volts increases and the terminal p.d. of the battery decreases. This means that the p.d. across each lamp will now be less than $10.0\,\text{V}$ and they will not light with normal brightness. **(3)**

33. Density and flotation

1 (a) $V = (0.05)^3 = 0.000125\,\text{cm}^3$. $\rho = \dfrac{m}{V} = \dfrac{0.100}{0.000125} = 800\,\text{kg m}^{-3}$ **(2)**

(b) The block must displace 100 g or $100\,\text{cm}^3$ of water, which would require the block sinking to a depth of $4.0\,\text{cm}$ or 80% of the depth of the block. **(2)**

2 (a) $T = mg = 1.00 \times 9.81 = 9.81\,\text{N}$ **(2)**

(b) $T = (1.00 - 0.125) \times 9.81 = 8.58\,\text{N}$ **(2)**

(c) The tension decreases because water is displaced and the mass experiences an upthrust that reduces the tension. It is reduced by the weight of 125 g of water $= 0.125 \times 9.81 = 1.23\,\text{N}$. **(2)**

(d) 1500 g **(2)**

34. Viscous drag

1 (a) Rearranging the formula gives $\eta = \dfrac{F}{6\pi r v}$. The 6π can be ignored as it has no units, so the units of viscosity are $\text{N m}^{-1}(\text{m s}^{-1})^{-1} = \text{N m}^{-2}\,\text{s} = \text{Pa s}$. **(2)**

(b)

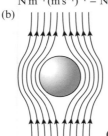

(3)

2 (a) Average time $= \dfrac{10.30 + 10.22 + 10.25 + 10.38 + 10.36}{5}$
$= 10.30\,\text{s}$. The average terminal velocity $= \dfrac{0.60}{10.30}$
$= 0.058\,\text{m s}^{-1}$. **(2)**

(b)

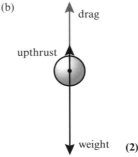

weight **(2)**

(c) $\eta = \dfrac{2 \times (7850 - 920) \times 9.81 \times (0.5 \times 10^{-3})^2}{9 \times 0.058} = 0.065\,\text{Pa s}$ **(2)**

35. Hooke's law

1 Hooke's law states that the force needed to extend a spring is proportional to the extension of the spring until it reaches its limit of proportionality. **(2)**

2 (a) load = 7.0 N, extension = 28 mm **(2)**

(b) The force constant k of a spring is defined by the equation $F = kx$, so $k = \dfrac{F}{x}$ or the slope of the straight part of the graph. $\dfrac{F}{x} = \dfrac{7.0}{0.028} = 250\,\text{N m}^{-1}$. **(2)**

(c) $E = \frac{1}{2}kx^2 = 0.5 \times 250 \times 0.020^2 = 0.05\,\text{J}$ **(2)**

3 (a) $W = Fs = 26 \times 0.080 = 2.1\,\text{J}$ **(2)**

(b) It becomes heat energy that is lost to the surroundings. **(2)**

36. Young modulus

1 (a) Tensile stress is the (tensile) force per unit area within the material. It is therefore defined in a similar way to pressure $= \dfrac{\text{force}}{\text{area}}$ and hence has the same units. **(2)**

(b) Tensile strain is the ratio of extension to original length. It is therefore the ratio of two lengths and has no units. **(2)**

(c) Young modulus is defined as stress divided by strain, and as strain has no units, the Young modulus and stress must have the same units. **(1)**

2 (a) tensile stress $\sigma = \dfrac{F}{A} = \dfrac{10}{\pi \times (0.25 \times 10^{-3})^2} = 51\,\text{MPa}$ **(2)**

(b) $\Delta l = \dfrac{\sigma l}{E} = \dfrac{51 \times 10^6 \times 3.0}{120 \times 10^9} = 1.3\,\text{mm}$ **(2)**

(c) $F = \sigma A = 70 \times 10^6 \times \pi \times (0.25 \times 10^{-3})^2 = 14\,\text{N}$ (or simply $\frac{70}{51} \times 10 = 14\,\text{N}$) **(2)**

(d) Once the wire exceeds the yield point and deforms plastically, it is no longer possible to use the Young modulus to predict extension, as this can only be used when the wire is behaving in a linear elastic manner. **(2)**

37. Exam skills 4 Stress, strain and the Young modulus

1 (a) If a single wire were measured and its temperature were to increase, it would increase in length, leading to an error. When two wires are used, both wires expand together, and no temperature-related extension is recorded even though both wires expand. **(2)**

(b)

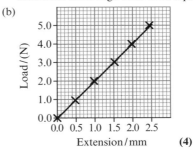

Extension / mm **(4)**

(c) Gradient $= \dfrac{5.00}{2.46} = 2.03\,\text{kg mm}^{-1}$ **(2)**

(d) Radius $= 0.30\,\text{mm} = 0.00030\,\text{m}$. Cross-sectional area $= \pi \times 0.00030^2 = 2.8 \times 10^{-7}\,\text{m}^2$ **(2)**

(e) Young modulus $= \dfrac{\sigma}{\varepsilon} = \dfrac{F}{A} \times \dfrac{l}{x} = \dfrac{mg}{A} \times \dfrac{l}{x}$.

The gradient is $\dfrac{m}{x}$ so

$$E = \dfrac{2.03 \times 10^3 \times 9.81 \times 3.00}{2.8 \times 10^{-7}} = 2.1 \times 10^{11} = 210\,\text{GPa}. \text{ (2)}$$

38. Waves

1 Rearranging $v = f\lambda$ gives $\lambda = \dfrac{v}{f} = \dfrac{3.00 \times 10^8}{96.1 \times 10^6} = 3.12\,\text{m}$,

so the length should be $\dfrac{3.12}{4} = 0.78\,\text{m}$. (2)

2 (a) $f = \dfrac{v}{\lambda} = \dfrac{1540}{0.00044} = 3.5 \times 10^6\,\text{Hz (3.5 MHz)}$ (2)

(b) A very short wavelength will reduce diffraction effects and increase the sharpness and resolution of the ultrasound image. (2)

3 (a) Wavelength $\lambda = 8.6\,\text{mm} = 8.6 \times 10^{-3}\,\text{m}$ (1)

(b) Time period $T = 25\,\mu\text{s} = 2.5 \times 10^{-5}\,\text{s}$. Frequency $f = \dfrac{1}{T}$

$= \dfrac{1}{2.5 \times 10^{-5}} = 4.0 \times 10^4\,\text{Hz}$ (2)

(c) Speed $v = f\lambda = 4.0 \times 10^4 \times 8.6 \times 10^{-3} = 340\,\text{m s}^{-1}$ (1)

39. Longitudinal and transverse waves

1 (a) (2), (b)

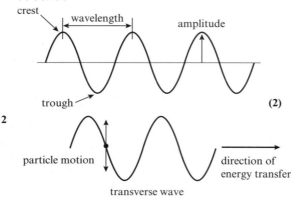

2

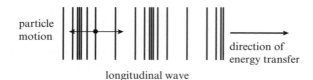

In the case of transverse waves, the particles carrying the wave oscillate at right angles to the direction of propagation of the wave or the direction of energy transfer. For longitudinal waves, the particles oscillate parallel to the direction that the wave is travelling. (In the case of electromagnetic waves, which are also transverse, the wave consists of oscillating electric and magnetic fields that are perpendicular to the direction of energy transfer.) (4)

3 C − ultrasound (1)

4 B − polarised (1)

5 C − minimum pressure (1)

40. Standing waves

1 (a) Microwaves incident on the metal sheet are reflected, creating two waves of similar amplitude but travelling in opposite directions. Where these waves are in phase, constructive interference results in the formation of antinodes, i.e. points of maximum amplitude, and where they are in antiphase, destructive interference results in nodes, i.e. points of zero or minimum amplitude. (4)

(b) When the probe encounters an antinode, a maximum output signal is produced because the wave amplitude is a maximum, whereas a node will produce a minimum output signal where the wave amplitude is a minimum. In addition, the maxima increase in amplitude as the probe moves toward the transmitter and the minima

become non-zero. This is because the reflected wave gets weaker and destructive interference is not complete, as the two amplitudes are no longer the same. (3)

(c) The distance between antinodes is half a wavelength, so the wavelength is 28 mm. (2)

41. Phase and phase difference

1 (a) The particles between two adjacent nodes in a standing wave always move in phase with each other. (1)

(b) The particles at two adjacent antinodes move in antiphase or are 180° or π radians out of phase with each other. (2)

2 A wavefront is a line joining points of equal phase to show the shape of a progressive wave. (1)

3 (a) Phase difference $= 2\pi \times \dfrac{x}{\lambda} = 2\pi \times \dfrac{4}{16} = \dfrac{\pi}{2}$ radians (see diagram)

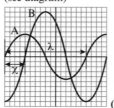

(2)

(b) $\dfrac{\pi}{2}$ radians = 90° (1)

4 (a) $\lambda = \dfrac{v}{f} = \dfrac{340}{850} = 0.40\,\text{m}$. The path difference is

BP − AP $= 0.20\,\text{m}$ or $\dfrac{\lambda}{2}$, so the phase difference at P will be π radians. (2)

(b) If the phase of A is reversed, i.e. shifted by π, a phase lead of π will become matched to give sound in phase with B. (1)

42. Superposition and interference

1 C − amplitudes must sum to zero (1)

2 A − the same frequency, f (1)

3 C − frequency, otherwise they will not remain in phase (1)

4

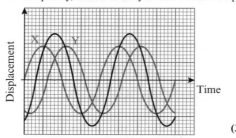

(3)

5 (a) Using Pythagoras' theorem: ACB $= 2 \times \sqrt{70.0^2 + 32.1^2}$
$= 154.0\,\text{cm}$ (2)

(b) Path difference $= 154.0 − 140.0 = 14.0\,\text{cm}$ (1)

(c) A path difference of one wavelength results in a phase difference of 2π. $\dfrac{14.0}{2.8} = 5$ wavelengths, so the phase difference is $5 \times 2\pi = 10\pi$ radians. (2)

(d) The path difference is a whole number of wavelengths, so constructive interference occurs at B. (1)

43. Velocity of transverse waves on strings

1 C − increasing the density of the metal, as this will increase the mass per unit length of the wire (1)

2 B − the string with the shortest length and greatest tension (1)

3 (a) 4650, 5540 (must be rounded to three significant figures) (1)

(b)

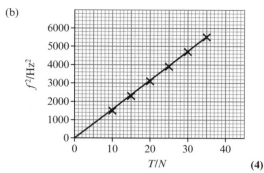

(4)

(c) The line of best fit is a straight line through the origin, so f^2 is directly proportional to T. **(2)**

44. The behaviour of waves at an interface

1 B and C **(2)**
2 (a) The time for one cycle of ultrasound is found using $T = \frac{1}{f}$. A pulse contains 10 cycles, so the duration is $10T = \frac{10}{40\,000} = 2.50 \times 10^{-4}$ s. **(1)**

(b) 20 pulses per second, so one is sent every $\frac{1}{20}$ or 0.050 s, so the total distance would be $s = vt = 340 \times 0.050 = 17$ m. **(2)**

(c) $\lambda = \frac{v}{f} = \frac{340}{40\,000} = 8.5 \times 10^{-3}$ m or 8.5 mm. **(1)**

(d) The above result suggests that if an object is farther away than $\frac{17}{2} = 8.5$ m, another pulse will be sent before the reflected pulse is received, which could 'confuse' the device. Also, the strength of the reflected signal will become very weak as the distance increases because the beam of ultrasound will spread out. **(2)**

45. Refraction of light and intensity of radiation

1 (a) $f = \frac{v}{\lambda} = \frac{3.00 \times 10^8}{532 \times 10^{-9}} = 5.64 \times 10^{14}$ Hz **(2)**

(b)

	In air	In water
Speed / m s^{-1}	3.00×10^8	2.26×10^8
Wavelength / nm	532	400
Frequency / Hz	5.64×10^{14}	5.64×10^{14}

(2)

2 The area of a sphere of radius r is given by $A = 4\pi r^2$ and intensity is power divided by area; the light spreads out over the surface of a sphere. The light output is 7.5% of 100 W, which is 7.5 W, so the intensity is $I = \frac{P}{A} = \frac{7.5}{4 \times \pi \times 2.0^2} = 0.15$ W m^{-2}. **(3)**

3

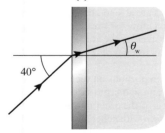

Diagram NOT accurately drawn.

(a) As $n \sin \theta = $ constant, at the air−glass boundary, we can say that $n_{air} \sin 40° = n_{glass} \sin \theta_g$, so $\sin \theta_g = \frac{n_{air} \sin 40°}{n_{glass}}$
$= \frac{1.00 \times 0.643}{1.52} = 0.423$ and $\theta_g = 25°$. **(2)**

(b) At the glass−water boundary $n_{glass} \sin \theta_g = n_{water} \sin \theta_w$,
so $\sin \theta_w = \frac{n_{glass} \sin \theta_g}{n_{water}} = \frac{1.52 \times 0.423}{1.33} = 0.483$ and
$\theta_w = 29°$. **(2)**

46. Total internal reflection

1 (a) The refractive index, n, is equal to the ratio of the speed of light in a vacuum to the speed of light in the material: $n = \frac{c}{v}$. (It is a measure of the amount of refraction caused by the material.) **(1)**

(b) The critical angle is the angle of incidence at which a light ray passing from one medium into another in which it travels faster will be refracted at 90° to the normal and will travel along the boundary between the media. (If the angle of incidence is larger than the critical angle, no light can exit the medium and total internal reflection occurs.) For a boundary between a medium with refractive index n and air, the critical angle C is defined by $\sin C = \frac{1}{n}$. **(3)**

(c) For total internal reflection to occur, the angle of incidence, which is 45°, must be greater than the critical angle. As $\sin C = \frac{1}{n}$, $C = \sin^{-1}\left(\frac{1}{1.49}\right) = 42°$, which is less than 45°, so total internal reflection can occur. **(2)**

2

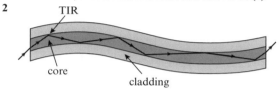

An optical fibre consists of an inner core of optically dense glass surrounded by a cladding layer of less optically dense glass. If a ray of light is incident at the core−cladding boundary at an angle greater than the critical angle, it will be totally internally reflected and transmitted down the fibre. **(4)**

47. Exam skills 5 Waves and their properties

1 (a) $n = \frac{\sin i}{\sin r}$, so $r = \sin^{-1}\left(\frac{\sin 45°}{1.5}\right) = 28°$ **(2)**

(b) When a ray of light meets a boundary between two different media, some light will pass across the boundary, undergoing refraction, and some will be reflected. When the ray passes into a medium in which it travels faster – for example from glass to air – the ray bends away from the normal. At a certain angle of incidence, the refracted ray will be bent at 90° to the normal and will travel along the boundary. This is the critical angle. If the angle of incidence is increased further, no light can exit the medium and total internal reflection occurs. **(3)**

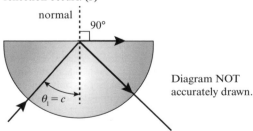

Diagram NOT accurately drawn.

(c) $n = \frac{1}{\sin C}$, so $C = \sin^{-1}\left(\frac{1}{1.5}\right) = 42°$ **(2)**

(d) The angle of incidence at the sloping surface is 28° + 31° = 59°, which is greater than the critical angle, so total internal reflection occurs. **(2)**

(e)

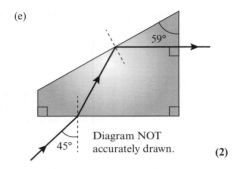

Diagram NOT accurately drawn. **(2)**

48. Lenses and ray diagrams

1 (a) (i)

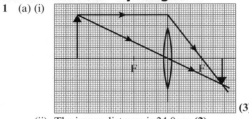

(3)

(ii) The image distance is 24.0 cm **(2)**

(iii) The linear magnification is $m = \dfrac{h_i}{h_o} = \dfrac{12.0}{20.0} = 0.60$

$\left(\text{or } m = \dfrac{v}{u} = \dfrac{24.0}{40.0} = 0.60\right)$. **(1)**

(iv) The image is real, inverted and diminished. **(2)**

(b) (i)

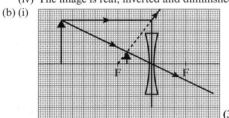

(3)

(ii) The image is virtual, upright and diminished. **(2)**

49. Lens formulae

1 (a) The focal length $f = 40.0$ cm $= 0.400$ m. $P = \dfrac{1}{f} = 2.50$ D
(2)

(b) $P = P_A + P_B = 2.50 + \dfrac{1}{0.500} = 4.50$ D **(1)**

(c) $f = \dfrac{1}{P} = \dfrac{1}{4.50} = 0.222$ m or 22.2 cm **(1)**

2 (a) Using $\dfrac{1}{u} + \dfrac{1}{v} = \dfrac{1}{f}$ gives $\dfrac{1}{25} + \dfrac{1}{v} = \dfrac{1}{10}$, so $\dfrac{1}{v} = \dfrac{1}{10} - \dfrac{1}{25}$

$= \dfrac{3}{50}$ and $v = 16.7$ cm **(2)**

(b) Using $\dfrac{1}{u} + \dfrac{1}{v} = \dfrac{1}{f}$ gives $\dfrac{1}{5} + \dfrac{1}{v} = \dfrac{1}{10}$, so $\dfrac{1}{v} = \dfrac{1}{10} - \dfrac{1}{5}$

$= -\dfrac{1}{10}$ and $v = -10.0$ cm **(2)**

(c) $m = \dfrac{v}{u} = \dfrac{10.0}{5.0} = 2.0$ **(1)**

(d) The image distance is negative, which means that the image is on the same side of the lens as the object and must be virtual, as rays only converge to form a real image on the opposite side of a converging lens. **(2)**

50. Plane polarisation

1 C – Ultrasound is a longitudinal wave. **(1)**

2 (a) The direction of polarisation of light is actually determined by a vector quantity. Unpolarised light oscillates in every direction but can be resolved into two perpendicular components, e.g. vertical and horizontal. 50% will be vertically polarised and 50% will be horizontally polarised. A vertical polarising filter will only let through the vertically polarised component, i.e. 50% of the light. **(2)**

(b) Light passing through the first filter will be plane-polarised in a direction at 45° to the second filter. This light can be resolved into components both parallel to and perpendicular to the second filter and the parallel component can be transmitted. **(2)**

(c) Light passing through the first filter will be plane-polarised in a direction perpendicular to the second filter and thus has zero component parallel to the second filter, so no light can be transmitted. **(2)**

3

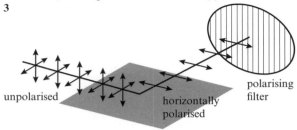

The reflected light is polarised horizontally. The lenses in the sunglasses will only let through light that is vertically polarised, so the reflected light will be blocked. Vertically polarised light will be unaffected. **(3)**

51. Diffraction and Huygens' construction

1 (a) Huygens' principle states that every point on a wavefront of a progressive wave can be considered to be a source of circular waves moving forward from that point. The resultant wavefront is found by the superposition of these circular waves. **(2)**

(b) By considering points along the wavefront as the wave emerges from the slit, and by constructing secondary 'wavelets' and then considering the result of the superposition of the latter, it is possible to construct the new wavefront of the diffracted wave.

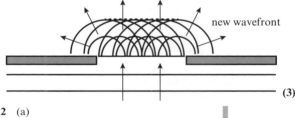

(3)

2 (a)

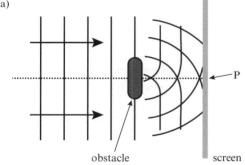

Light is diffracted by the edges of the obstacle such that waves are able to arrive at P, as seen in the diagram. Diffracted waves from the top and bottom of the obstacle arrive at the centre point P in phase, so constructive interference occurs and bright light is seen. **(3)**

(b) Diffraction phenomena such as that in (a) above only occur to any significant extent when the object or obstacle is of the order of a few wavelengths in size and are best seen with coherent sources of light. For this reason, we normally just see shadows for large obstacles and everyday light sources. **(2)**

52. Using a diffraction grating to measure the wavelength of light

1 (a) 250 lines mm^{-1} means there are 2.50×10^5 lines m^{-1} and

$d = \dfrac{1}{2.50 \times 10^5} = 4.00 \times 10^{-6}$ m. **(1)**

(b) $\theta = \tan^{-1}\left(\dfrac{0.319 + 0.321}{2 \times 2.00}\right) = 9.09°$ **(2)**

(c) $\lambda = \dfrac{d \sin \theta}{n}$ and $n = 1$

$\lambda = 4.00 \times 10^{-6} \times \sin 9.09° = 6.32 \times 10^{-7}\,\text{m}$ (632 nm) **(2)**

(d) For $n = 2$, $\theta = \tan^{-1}\left(\dfrac{0.668 + 0.671}{2 \times 2.00}\right) = 18.5°$

$\lambda = \dfrac{d \sin \theta}{n} = \dfrac{4.00 \times 10^{-6} \times \sin 18.5°}{2} = 6.35 \times 10^{-7}\,\text{m}$

so mean wavelength (for first and second orders)
$\lambda = 6.34 \times 10^{-7}\,\text{m}$ (634 nm) **(3)**

(e) You could increase the length l to decrease the percentage error in both x and l, and you could also use the higher orders and find the mean wavelength from a larger amount of data. **(2)**

53. Electron diffraction

1 (a) Diffraction occurs due to the superposition of waves. Therefore, if we see diffraction patterns produced by electrons passing through graphite, we must conclude that the electrons are behaving as waves. **(2)**

(b) Diffraction only occurs to an appreciable extent when the wavelength of the wave being diffracted is comparable to the spacing in the object doing the diffracting. Here, electron diffraction occurs because the wavelength of the electrons is comparable to the atomic spacing in the graphite crystals. **(2)**

(c) $D - \dfrac{p^2}{2m} = \dfrac{mv^2}{2}$ **(1)**

(d) The electrons of charge e are accelerated by a p.d. V and gain kinetic energy equal to eV. This means that eV

$= \tfrac{1}{2}mv^2$ and $v = \sqrt{\dfrac{2eV}{m}} = \sqrt{\dfrac{2 \times 1.60 \times 10^{-19} \times 3000}{9.11 \times 10^{-31}}}$

$= 3.25 \times 10^7\,\text{m s}^{-1} \approx 3 \times 10^7\,\text{m s}^{-1}$. **(2)**

(e) The de Broglie wavelength of a particle is given by $\lambda = h / mv$ where h is the Planck constant, 6.63×10^{-34} J s.

$\lambda = \dfrac{h}{mv} = \dfrac{6.63 \times 10^{-34}}{9.11 \times 10^{-31} \times 3.25 \times 10^7}$

$= 2.24 \times 10^{-11}\,\text{m}$ **(1)**

(f) A − of smaller radius and brighter **(1)**

54. Waves and particles

1 (a) At one time, there were competing theories: some scientists (for example, Huygens) considered that light propagated through space as a wave, whereas others (most notably Newton) believed that light propagated through space as a stream of particles. However, the earlier particle model of light could not explain phenomena such as diffraction. Quantum theory proposed that light consisted of quanta of energy or photons, as suggested by the photoelectric effect. At the same time, it was also recognised that light must nevertheless still behave like a wave under some circumstances, leading to the idea that light is neither just a wave nor just a particle phenomenon, but has aspects of both. **(3)**

(b) Young's double slit experiment provides evidence of the wave nature of light. Only waves exhibit superposition effects, such as diffraction and interference, both of which are involved in the formation of fringes when light passes through a double slit. It is therefore appropriate to conclude from the experiment that light travels as waves. **(3)**

(c) An experiment to show the photoelectric effect using a gold-leaf electroscope and zinc plate demonstrates the particle nature of light. The ability of light to discharge a charged gold-leaf electroscope depends on its frequency, in that only high-frequency light can discharge the gold-leaf electroscope. There is no explanation for this observation in a wave theory, so it must be concluded that light travels as discrete particles or quanta. **(3)**

2 The de Broglie wavelength of a particle is given by

$\lambda = \dfrac{h}{p} = \dfrac{h}{mv} = \dfrac{6.63 \times 10^{-34}}{9.11 \times 10^{-31} \times 3.00 \times 10^7} = 2.4 \times 10^{-11}\,\text{m}$ **(2)**

55. The photoelectric effect

1 (a) Electrons would not be emitted as photoelectrons if the zinc plate were positively charged or neutral, as a positively charged plate would attract the negatively charged photoelectrons and prevent emission. **(2)**

(b) The frequencies in visible light are lower than the threshold frequency for zinc; that is, electrons do not gain enough energy from the photons to escape from the surface, so no emission of photoelectrons can take place. **(2)**

(c) Ultraviolet light has a higher frequency than visible light. Its frequency is greater than the threshold frequency for zinc and so it does provide enough energy to eject photoelectrons and discharge the gold-leaf electroscope. **(2)**

(d) Different metals require greater or lesser amounts of energy in order to eject photoelectrons, so different metals have different threshold frequencies. **(2)**

2 (a) The work function of a metal is the minimum amount of energy required to eject an electron from the surface of the metal. **(1)**

(b) $\phi = hf_0$, so $\dfrac{\phi}{h} = f_0 = \dfrac{2.36 \times 1.60 \times 10^{-19}}{6.63 \times 10^{-34}} = 5.70 \times 10^{14}\,\text{Hz}$ **(2)**

(c) The maximum KE is the photon energy minus the work function $= hf - \phi = h(f - f_0) = 6.63 \times 10^{-34}$

$\times \left(\dfrac{3.00 \times 10^8}{465 \times 10^{-9}} - 5.70 \times 10^{14}\right) = 4.98 \times 10^{-20}\,\text{J}$ **(2)**

56. Line spectra and the eV

1 Electrons in atoms occupy certain discrete energy levels. If an electron gains energy – for example, thermal energy – it can be promoted to a higher energy level. It can then lose this energy when it moves back to a lower energy state. The energy that is lost is emitted as a photon of electromagnetic radiation, including visible light. As moving between energy levels involves a fixed amount of energy, only certain photon energies will be produced. Hence, the spectrum consists of lines corresponding to certain frequencies as $E = hf$. **(4)**

2 (a) -0.850 eV and -0.544 eV (top) **(1)**

(b) $-13.6 \times 1.60 \times 10^{-19} = 2.18 \times 10^{-18}$ J **(1)**

(c) $\Delta E = -1.51, - (-3.40) = 1.89$ eV
$1.89 \times 1.60 \times 10^{-19} = 3.02 \times 10^{-19}$ J **(2)**

(d) $\Delta E = hf = \dfrac{hc}{\lambda}$, which gives $\lambda = \dfrac{hc}{\Delta E}$, where ΔE is the energy calculated in (c)

$\lambda = \dfrac{6.63 \times 10^{-34} \times 3.00 \times 10^8}{3.02 \times 10^{-19}}$

$= 6.58 \times 10^{-7}\,\text{m}$ **(2)**

(e) Energy of photon $= \dfrac{hc}{\lambda}$

$= \dfrac{6.63 \times 10^{-34} \times 3.00 \times 10^8}{103 \times 10^{-9}} = 1.93 \times 10^{-18}\,\text{J} = 12.07\,\text{eV}$.

The difference in energy between $n = 1$ and $n = 3$ is $-1.51 - (-13.6) = 12.07$ eV or 1.93×10^{-18} J, so yes, the photon would be absorbed as its energy corresponds to the change in energy between those two levels. **(3)**

57. Exam skills 6 The quantum nature of light

1 (a) The threshold frequency is the lowest frequency of light that will result in the emission of photoelectrons from a particular metal surface. **(2)**

(b) **(1)**

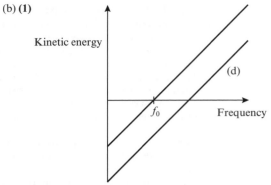

(c) The threshold frequency would be the frequency of photons that just have enough energy to knock electrons out of the metal surface but no more, i.e. when $E_{max} = 0$ and $hf_0 = \phi$ **(2)**

(d) See graph. **(2)**

(e) A photon is a concentrated packet of electromagnetic radiation, e.g. light or a quantum of energy of electromagnetic radiation. Its energy is given by $E = hf$. **(2)**

(f) All the energy of a photon is acquired by one particular photoelectron, but part of that energy is the work function, the minimum energy required to escape from the metal surface. If all the remaining energy is transferred into movement of the photoelectron, i.e. kinetic energy, it is the maximum, but in fact some of the energy might be transferred somewhere else, e.g. to thermal energy in the metal.**(2)**

58. Impulse and change of momentum

1 (a) impulse; N s. **(2)**

(b) 2 kg m s^{-1} (area under graph) **(3)**

(c) $\Delta p = mv - (-mv) = 2mv$, so $v = \dfrac{\Delta p}{2m}$

However, this is the total change of momentum as the ball reverses direction. The collision is elastic, so kinetic energy is conserved, therefore velocity just before landing = 5.0 m s^{-1} **(3)**

2 • Falling child has a rapid change of momentum (to zero) on impact.
  • Compression of surface increases the time over which the momentum change occurs.
  • Force is rate of change of momentum.
  • Force is reduced, so injury is likely to be less severe. **(4)**

59. Conservation of momentum in two dimensions

1 (a) The total momentum of a closed system is conserved (constant).
A closed system is one that has no external resultant forces acting on it. Resultant forces would cause the momentum to change. **(3)**

(b) (i) 6.0×10^7 N s or kg m s^{-1} **(2)**

(ii)

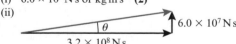

Initial momentum = $mv = 10^4 \times 3.2 \times 10^4$ kg m s^{-1}

$\sin\theta = \dfrac{6.0 \times 10^7}{3.2 \times 10^8}$

$\theta = 11°$ **(3)**

(iii) The total momentum of ejected gases is equal in magnitude to the change of momentum of the rocket (6.0×10^7 N s) but opposite in direction to this change. **(2)**

2 For example;

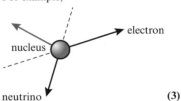

(3)

Momentum is always conserved, so the momentum vector for the recoil must be equal and opposite to the vector sum of the other two momentum vectors. In practice, this means the direction of recoil must be somewhere between the two dotted lines above.

> If the arrows represent the momentum vectors, showing the magnitude as well as the direction of the momenta, then the recoil momentum will have the same magnitude as the sum of the two vectors, but the opposite direction.

60. Elastic and inelastic collisions

1 (a) The total energy before will be equal to the total energy after. The initial kinetic energy of the man will be partially transferred to the boat, but some will be transferred to other forms (e.g. thermal energy by friction) as the man jumps into the boat and stops. The linear momentum of the man will be equal to the total linear momentum of the man and boat after the collision, as the momentum is conserved. **(4)**

(b) By conservation of momentum, velocity of man and boat $v_2 = \dfrac{m_1 v_1}{m_2} = \dfrac{560}{300}$

$= 1.87$ m s$^{-1} \approx 1.9$ m s^{-1} **(3)**

(c) (i) KE of man = $\frac{1}{2}mv^2 = 2240$ J; KE of man and boat after collision = 523 J **(2)**

(ii) KE has not been conserved so it is an inelastic collision. **(2)**

(iii) The boat's KE has been transferred to the water as KE and by heating. Its momentum has been transferred to the water. **(2)**

61. Investigating momentum change

1 (a) The tilt is to compensate for friction. If the student did not do this the resultant force on the trolley would be less than the weight of the masses and hanger, so their calculated values for the impulse would be too large. **(3)**

(b) By changing the number of masses on the mass hanger. The resultant force is mg, where m is the total mass of hanger and attached masses. **(2)**

(c) This is to keep the total mass of the system (trolley + masses + hanger) constant. Total mass is a control variable. **(2)**

(d) Impulse = Ft, so she must multiply the weight of the hanger + masses by the time for the trolley to move between the two light gates. This time is recorded by a datalogger. **(2)**

(e) She needs to multiply the total mass of the system (trolley + hanger + masses) by the final velocity. The final velocity is recorded by the datalogger when the trolley passes through the second light gate. **(2)**

(f) Momentum change is equal to impulse, so the graph should be a straight line through the origin with gradient equal to one. **(3)**

62. Exam skills 7 Impulse and change of momentum

1 (a)

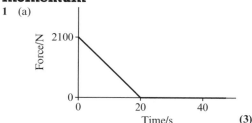

(3)

(b) Impulse equals area under graph of resultant force over time $= \frac{1}{2} \times 20 \times 2100 = 21\,000$ N s **(2)**

(c) $\Delta v = \frac{\Delta p}{m} = \frac{21\,000}{1400} = 15$ m s^{-1} **(2)**

(d) Increasing from zero with decreasing gradient. Gradient becomes zero at 20 s and velocity of 15 m s^{-1}. Velocity remains constant at 15 m s^{-1} for remaining time. **(3)**

(e) Displacement can be estimated from the velocity–time graph and is greater than 450 m. $W = Fs$, which will be greater than $2100 \times 450 = 945\,000$ J. **(3)**

(f) KE $= \frac{\Delta p^2}{2m} = \frac{21\,000^2}{2 \times 1400} = 158$ kJ, so work against frictional forces = answer to (e) − 158 = 787 kJ. **(3)**

63. Describing rotational motion

1 C

2 (a) $\frac{2\pi}{3.16 \times 10^7} = 1.99 \times 10^{-7}$ rad s^{-1} **(3)**

(b) 2.98×10^4 m s^{-1} **(1)**

(c) (i) 1.87 years **(3)**

(ii) $\frac{1}{1.87} \times 2\pi = 3.4$ radians, 193° **(3)**

64. Uniform circular motion

1 (a) During a short time the child will move from A to B and the velocity vector will change direction. Acceleration is defined as the rate of change of velocity, so the child must be accelerating. For small times, the change in velocity from A to B is a vector at 90° to the velocity at A and toward the centre of the circle, so the acceleration is also toward the centre of the circle. **(4)**

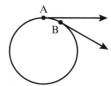

(b) $a = r\omega^2 = 2.74$ m s^{-2} **(2)**

(c) (i) From Newton's second law − resultant force causes acceleration. Here there is an acceleration, so there must be a resultant force in the same direction, toward the centre. **(2)**

(ii) Friction from the surface of the roundabout; the seat and hand rails on the roundabout. **(1)**

(d) 26.2 radians **(3)**

65. Centripetal force and acceleration

1 (a) $F = \frac{mv^2}{r} = 8640$ N toward the centre of the circle. **(2)**

(b) The car's wheels are turned and the tyres exert an outward frictional force on the road. By Newton's third law, the road produces an equal inward force on the car's tyres. **(2)**

(c) There is a limit to the maximum frictional force from the road. At higher speeds, the required centripetal force is greater $\left(F = \frac{mv^2}{r}\right)$. If this is greater than the maximum frictional force, then the car cannot turn in a circle of that radius at that speed and so it skids. **(2)**

2 (a) When the bob moves through its lowest position, it is moving in circular motion, so there must be a resultant force toward the centre of the circle (i.e. the point of suspension). The upward tension must therefore be greater than the downward weight in order to provide this resultant centripetal force. **(3)**

(b) Resultant force = tension − weight $= \frac{mv^2}{r}$

Tension $= mg + \frac{mv^2}{r} = 0.5886 + 0.11772 = 0.71$ N **(4)**

66. Electric field strength

1 (a) An electric field is a region of space in which electric charges experience a force.
Electric field strength E is the force per unit charge acting on a positive test charge at a point in space.
$E = \frac{F}{Q}$ (measured in N C^{-1} or V m^{-1}). **(2)**

(b) $V = 3 \times 10^6 \times 2.5 \times 10^{-3} = 7500$ V **(2)**

(c) (i) $E = \frac{V}{d} = \frac{500}{4 \times 10^{-3}} = 125$ kV m^{-1} **(1)**

(ii) $E = \frac{500}{8 \times 10^{-3}} = 62.5$ kV m^{-1} **(1)**

(iii)

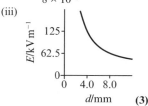

(3)

2

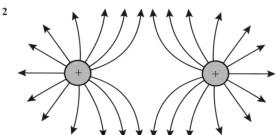

Essential points: field lines have correct direction indicated; no field lines cross; pattern is correct. **(3)**

67. Electric field and electric potential

1 (a) $W = qV = 1.6 \times 10^{-19} \times 420 = 6.72 \times 10^{-17}$ J **(2)**

(b) $W = $ final KE $= \frac{1}{2}mv^2$, so $v^2 = \frac{2W}{m} = \frac{2 \times 6.72 \times 10^{-17}}{9.1 \times 10^{-31}}$

$v = 1.2 \times 10^7$ m s^{-1} **(3)**

2 (a) $E = \frac{V}{d} = \frac{1 \times 10^9}{300} = 3.3 \times 10^6$ V m^{-1} **(2)**

(b) energy $= QV = 1.6 \times 10^{10}$ J
The energy is transferred into light and sound. It also heats the air and the ground where it strikes, and is transferred into chemical potential energy where it causes chemical reactions in the atmosphere. **(3)**

3 Electric field points toward negative charge.
Equipotentials are concentric circles centred on charge (i.e. perpendicular to field lines).
Separation of equipotentials increases away from charge.
Highest potential is farthest equipotential from charge. **(4)**

68. Forces between charges

1 C − because $F = \frac{Q_1 Q_2}{4\pi\varepsilon_0 r^2}$ **(1)**

2 (a) To the: right; left; right **(3)**

(b) Taking forces to the right as positive, the resultant force on the electron is (by Coulomb's law)

$F = \frac{Qe}{4\pi\varepsilon_0 r^2} - \left(\frac{-Qe}{4\pi\varepsilon_0 r^2}\right) = \frac{2 \times 8.0 \times 10^{-20} \times (-1.6 \times 10^{-19})}{4\pi\varepsilon_0 r^2}$

$= -3.7 \times 10^{-9}$ N (attraction). The negative sign indicates a force to the left. **(4)**

(c) Resultant force = force due to $-Q$ + force due to $+Q$:

$$F = \frac{-Qe}{4\pi\varepsilon_0 r^2} + \frac{Qe}{4\pi\varepsilon_0 (3r)^2} = 1.6 \times 10^{-9}\,\text{N to right (repulsion)}$$ **(4)**

(d) The molecules might align with the field, so that the entire material becomes a dipole. **(2)**

69. Field and potential for a point charge

1. (a) $V = \dfrac{Q}{4\pi\varepsilon_0 r}$

   $Q = V4\pi\varepsilon_0 r = 4.00 \times 10^5 \times 4\pi \times 8.85 \times 10^{-12} \times 0.2$
   $= 2.22 \times 10^{-6}\,\text{C}$ **(3)**

   (b) V at 0.20 m from dome surface is effectively V at 0.40 m from central point, so at $2r$.

   $V \propto \dfrac{1}{r}$: if r is doubled V is halved, so $V = 200\,\text{kV}$ **(2)**

   (c) $r = 0.25$, so potential at 0.05 m above dome is

   $\dfrac{400 \times 0.20}{0.25} = 320\,\text{kV}$ **(2)**

   (d) (i) Electric field strength is the negative potential gradient. $\left(E = \dfrac{Q}{4\pi\varepsilon_0 r^2} = \dfrac{V}{r}\right)$ **(1)**

   (ii) $E = \dfrac{V}{r} = \dfrac{4.00 \times 10^5}{0.20} = 2.00 \times 10^6\,\text{V m}^{-1}$ **(2)**

   (iii) E at 0.20 m from dome surface is effectively E at 0.40 m from central point, so at $2r$.

   $E \propto \dfrac{1}{r^2}$: if r is doubled E is divided by four,

   so $E = 5.00 \times 10^5\,\text{V m}^{-1}$ **(2)**

   (e) 400 kV. If the electric field is zero, the potential gradient is also zero, so there is no change in potential within the dome and all points have the same potential as the surface of the conductor. **(2)**

70. Capacitance

1. A **(1)**
2. (a) 2.64 mC **(2)**

   (b) 0.12 A through both ammeters. **(2)**

   (c) As the capacitor charges, the potential difference across it increases. This opposes the e.m.f. of the battery, so the potential difference across the resistor (which is the difference of these two values) falls. **(2)**

   (d) 80% charge $Q = 0.8 \times 2.64 \times 10^{-3} = 0.002112\,\text{C}$, so p.d.

   across capacitor $V = \dfrac{Q}{C} = 9.6\,\text{V}$

   Therefore, p.d. across resistor is $12 - 9.6 = 2.4\,\text{V}$
   Current in circuit when capacitor is 80% charged

   $I = \dfrac{V}{R} = \dfrac{2.4}{100} = 0.024\,\text{A}\;(24\,\text{mA})$. **(4)**

71. Energy stored by a capacitor

1. (a) $C = \dfrac{Q}{V} = 50\,\mu\text{F}$ **(2)**

   (b) (i) area under line. **(1)**

   (ii) $W = \frac{1}{2}QV = \frac{1}{2} \times 2000 \times 10^{-6} \times 40 = 4.0 \times 10^{-2}\,\text{J}$ **(2)**

   (iii) $1.0 \times 10^{-2}\,\text{J}$ **(2)**

2. (a) $C = \dfrac{Q}{V}$ so $Q = CV = 22 \times 10^{-6} \times 9.0 = 198 \times 10^{-6}\,\text{C}$ **(2)**

   (b) $W_b = QV = 198 \times 10^{-6} \times 9.0 = 1.78 \times 10^{-3}\,\text{J}$ **(2)**

   (c) $W = \frac{1}{2}QV = \frac{1}{2} \times 198 \times 10^{-6} \times 9.0 = 8.91 \times 10^{-4}\,\text{J}$ **(2)**

   (d) Some of the energy supplied by the battery is used to pass current through the resistor. This transfers exactly 50% to heat. The value of the resistance makes no difference to the work done by the battery or the energy stored on the capacitor (R does not feature in the calculation of either). **(4)**

72. Charging and discharging capacitors

1. C **(1)**
2. (a) $Q = CV = 3.6\,\text{mC}$ **(1)**

   (b)

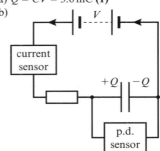

   (c) As the electrons carrying charge around the circuit accumulate on the negative plate and drain away from the positive plate of the capacitor, the p.d. across the capacitor increases. This reduces the p.d. across the resistor, so the rate of flow of charge in the circuit also falls. The rate at which the capacitor charges falls exponentially and finally levels off when the p.d. across the capacitor is equal but opposite to the e.m.f. of the battery. **(6)**

73. The time constant

1. (a) The time constant determines how long it takes for a capacitor to charge or discharge: over one time constant the charge falls to approximately 37% of its initial value. **(3)**

   (b) Units of RC are $\Omega\,\text{F} = \text{V A}^{-1}\,\text{C V}^{-1} = \text{A}^{-1}\,\text{C}$
   $= \text{C}^{-1}\,\text{s C} = \text{s}$ **(2)**

2. (a) $RC = 0.022\,\text{s}$ **(2)**

   (b) $(0.37)^5 = 0.007$, that is, in five time constants a discharged capacitor will be 99.3% charged, so $5 \times CR = 0.11\,\text{s}$ **(2)**

   (c) Increasing C increases the final charge ($Q = CV$), so it takes longer to supply this charge. Increasing R decreases the charging current, so again it takes longer to supply the charge. **(4)**

3. time constant $CR = 50 \times 10^{-6} \times 20 \times 10^3 = 1\,\text{s}$. $5CR = 5\,\text{s}$ **(2)**

74. Exponential decay of charge

1. (a) Charging curve rising to almost 6 V in about 2.5 s then constant at 6 V to 3 s. Discharging curve reaching almost zero at about 5.5 s and then remaining close to zero. **(6)**

   (b) $Q = CV = 6.0\,\text{mC}$ **(1)**

   (c) $I = \dfrac{V}{R} = 12\,\text{mA}$ at $t = 0$ **(2)**

   (d) $Q = Q_0\,e^{-t/RC}$ so when charge is halved,

   $\dfrac{Q}{Q_0} = e^{-t/RC} = 0.5$

   $\ln 0.5 = -t/RC$

   $t = RC \ln 2 = 0.35\,\text{s}$ after it starts to discharge (3.35 s on graph) **(3)**

   (e) Inverse of graph for capacitor, so that p.d. across R and p.d. across C add to 6.0 V at all times. **(2)**

75. Exam skills 8 Capacitors

1. (a) Time constant for the charging circuit is 4.0 s. Charging is 95% complete after three time constants, so the voltage across the capacitor is well above 60 V and the flash will work. **(3)**

   (b) (i) $W = \frac{1}{2}CV^2 = 0.014\,\text{J}$ **(2)**

   (ii) $W = 0.0036\,\text{J}$ **(1)**

   (c) $\dfrac{V}{V_0} = e^{-t/RC} = 0.5$ so $t/RC = \ln 2$ and $t = 1.4\,\text{ms}$ **(2)**

   (d) If voltage falls from 120 V to 60 V, energy stored falls from 14 mJ to 3.6 mJ ie. a fall of 10.4 mJ in 1.4 ms. So

   power $= \dfrac{\text{energy}}{\text{time}} = \dfrac{10.4\,\text{mJ}}{1.4\,\text{ms}} = 7.4\,\text{W}$. **(4)**

   (e) The flash lamp stops conducting when the p.d. across it falls below 60 V, so the supply only has to recharge the capacitor from 60 V to 120 V. This takes less time than charging from 0 V. **(2)**

76. Describing magnetic fields

1 D (1)

2 A (1)

3 (a) $\phi = BA \sin \theta = 36 \times 10^{-6} \times 2.0 \times \sin 30° = 3.6 \times 10^{-5}$ Wb
(3)

(b) $B_{horizontal} = 36 \times 10^{-6} \times \cos 30° = 31.2 \times 10^{-6}$ T
$NB_h A = 500 \times 31.2 \times 10^{-6} \times 0.25 = 3.9 \times 10^{-3}$ Wb.turns
(3)

4 Uniform magnetic field: constant magnitude and direction at all points − parallel equally spaced field lines. The Earth's field is not uniform on the large scale, but over distances small compared with the radius of the Earth it is approximately uniform. (4)

77. Forces on moving charges in a magnetic field

1 (a) Arc of a circle with centre of curvature below point of entry to field. (2)

(b) As in (a) but larger radius of curvature ($r = \frac{mv}{Bq}$ and v is larger). (2)

(c) The electron will be undeflected. The force on a moving charge results from the perpendicular component of the magnetic field to the velocity of the particle. If the particle moves along the field line, the perpendicular component is zero. (2)

2 (a) $r_1 = \frac{238 \, u \times v}{Bq}$ and $r_2 = \frac{235 \, u \times v}{Bq}$

$2 \times (r_1 - r_2) = \frac{2 \times (238 - 235) \, u \times v}{Bq} = \frac{6u \times v}{Bq}$ (4)

(b) Decrease B or increase the initial speed of the ions. (2)

78. Electromagnetic induction − relative motion

1 (a) (i) 0 Wb.turns

(ii) $N\Phi = NBA = 2.4 \times 10^{-3}$ Wb.turns

(iii) 0 Wb.turns

(iv) -2.4×10^{-3} Wb.turns (4)

(b) y-axis scaled -2.5×10^{-3} to $+2.5 \times 10^{-3}$ and unit is Wb.turns. Shape is a sine curve with one cycle up to $t = T$. (4)

(c) (i) Shape is a negative cosine with one cycle up to $t = T$. (2)

(ii) Faraday's law says that induced e.m.f. is proportional to negative rate of change of flux linkage, so the values on the e.m.f. graph are the negative gradients of the flux linkage graph at the same time. (3)

79. Changing flux linkage

1 (a) When the switch is closed, the current in the primary coil increases from zero. This causes an increase in the magnetic flux in the iron core.
This changing flux links the secondary coil and induces an e.m.f. across it (Faraday's law).
The secondary coil is part of a complete circuit, so a current flows and creates a magnetic field in the coil surrounding the compass. The compass deflects to align with this field.
Once the current has reached a steady value, the flux linkage is no longer changing, so the induced e.m.f. falls to zero and so does the induced current. The compass needle now returns to its original position. (8)

(b) The iron core increases the magnetic flux and provides a link between the two coils. (2)

(c) As the switch is opened, the current falls rapidly to zero, so the flux linkage rapidly falls to zero too. This induces an e.m.f. across the secondary coil in the opposite direction to the one in (a). A momentary current flows and the compass needle flicks to the right and then returns to its original position. (4)

(d) A.C. currents are continually changing, so they create a continually changing magnetic flux in the core.
This induces a continually changing e.m.f. across the

secondary coil of the transformer. If D.C. were used, there would be only a pulse of induced e.m.f. whenever the current was switched on or off. (3)

80. Faraday's and Lenz's laws

1 (a) • As the magnet moves up and down, its magnetic field creates a changing flux through the coil.
• Induced e.m.f. in the coil is directly proportional to rate of change of flux linkage (Faraday's law).
• The switch is open, so there is not a complete circuit − no current can flow. (3)

(b) • When the switch is closed, the induced e.m.f. will cause current to flow in the coil.
• The direction of current flow will oppose the change that caused it (Lenz's law).
• The currents will make the coil into an electromagnet, which repels the approaching magnet and attracts the receding magnet, damping the oscillations. (3)

(c) The moving magnet has to do work against the forces from the coil. This transfers mechanical energy to electrical energy in the coil. Energy is conserved. (2)

2 (a) change in flux linkage = $NB\Delta A$
$= 50 \times 20 \times 10^{-3} \times 25 \times 10^{-4} = 2.5 \times 10^{-3}$ Wb.turns (2)

(b) $\varepsilon = -\frac{d(N\Phi)}{dt}$, so average $\varepsilon = \frac{2.5 \times 10^{-3}}{0.20} = 0.0125$ V (3)

81. Alternating currents

1 (a) D.C. flows continuously in one direction around the circuit.
For A.C. the direction of current flow changes frequently. (2)

(b) peak voltage = 10 V, r.m.s. voltage = $\frac{V_{peak}}{\sqrt{2}} = 7.1$ V,
time period = 20 ms,
frequency = 50 Hz (4)

(c) (i) $\frac{V_{rms}}{R} = I_{rms} = 0.071$ A (1)

(ii) Average power = $V_{rms}I_{rms} = 0.50$ W (2)

(d) (i) The lamp will be dimmer. The power must be calculated using r.m.s. values, and the r.m.s. value of the supply voltage is less than 9 V. (3)

(ii) Zero power at 5 ms & 15 ms since, for a purely resistive load, power is max when V is max (at 0 ms, 10 ms & 20 ms). (3)

82. Exam skills 9 Electromagnetic fields

1 (a) 50 Hz (2)

(b) Average power = $I_{rms}^2 R = \frac{I_{peak}^2 R}{2} = \frac{(0.020)^2 \times 5.0}{2}$ (3)
$= 0.001 = 1.0$ mW (3)

(c) • Current in the first coil creates a magnetic field through the coil.
• The current is alternating, so it creates an alternating flux.
• The alternating magnetic flux links the second coil.
• By Faraday's law, there is an induced e.m.f. proportional to the rate of change of magnetic flux linkage through the second coil. (4)

(d) No energy is transferred. The second coil is an open circuit, so no current flows. Power requires both e.m.f. and current ($P = IV$). (2)

83. The Rutherford scattering experiment

1 (a) In a vacuum, the alpha particles will not be deflected or stopped by colliding with air molecules. (1)

(b) Gold can be beaten into a very thin foil. (1)

(c) Most of the alpha particles passed through the gold foil with little or no deflection, even though the foil was hundreds of atoms thick. (2)

(d) (i) Alpha particles are known to carry a positive charge. A force is required to deflect them. Rutherford assumed that this force was an electrostatic force between the positive charge of the alpha particle and a small volume of concentrated charge − the nucleus. (2)

(ii) The proportion of alpha particles scattered through large angles was very small. This suggested that the chance of passing close to a nucleus was very low, so the nucleus must occupy a very small fraction of the space inside the atom. **(2)**

(e) Lowest alpha particle approaches closest to nucleus and is then deflected back along same path.
Middle alpha particle deflects upwards.
Top alpha particle deflects least and crosses path of middle alpha particle. **(3)**

(f) Since the nucleus is much smaller than the atom and yet contains most of its mass, it must have a much higher density than ordinary matter. **(2)**

(g) With a lower atomic mass there will be a lower positive charge on the nucleus of tungsten and so it will exert a smaller force on alpha particles than a gold nucleus at the same distance. This means that alpha particles can get closer to the tungsten nucleus (smaller closest approach) and the proportion deflected through large angles will be smaller than for gold. **(3)**

84. Nuclear notation

1 (a)

Nuclide	Number of protons	Number of neutrons	Number of electrons
$^{7}_{3}$Li	3	4	3
$^{108}_{47}$Ag	47	61	47
$^{222}_{86}$Rn	86	136	86
$^{244}_{94}$Pu	94	150	94	**(12)**

(b) (i) Helium nucleus – 2 protons and 2 neutrons **(3)**
(ii) $^{222}_{86}$Rn → $^{218}_{84}$Po + $^{4}_{2}$He **(3)**
(iii) The totals are unchanged.
charge: 86 = 84 + 2; nucleon number: 222 = 218 + 4 **(2)**

2 The key result from Rutherford's experiment is that a very small number of alpha particles are deflected through large angles or back-scattered. The 'plum pudding' model cannot explain this observation, because the positive charge in the model is distributed evenly through the entire volume of the atom. This would result in all the alpha particles being deflected only a little or not at all. The positive charge and most of the atomic mass must therefore be concentrated in one small region inside the atom, unlike the 'plum pudding' model. **(4)**

85. Electron guns and linear accelerators

1 (a) (i) The low-voltage supply supplies the energy to heat the cathode, so that electrons are emitted by thermionic emission. **(2)**
(ii) The high-voltage supply supplies the energy to accelerate the electrons in the electron gun. **(2)**
(iii) There must be a vacuum inside the tube, otherwise the electrons would be deflected by collisions with air molecules and the beam could not be formed. **(2)**

(b) (i) KE = $\frac{1}{2}mv^2 = eV = 1.6 \times 10^{-19} \times 2500 = 4.0 \times 10^{-16}$ J **(2)**
(ii) KE = $\frac{1}{2}mv^2 = 4.0 \times 10^{-16}$ J
$$v = \sqrt{\frac{2 \times 4.0 \times 10^{-16}}{9.1 \times 10^{-31}}} = \sqrt{8.79 \times 10^{14}}$$
$$= 3.0 \times 10^7 \, \text{ms}^{-1} \, \textbf{(2)}$$
(iii) The two deflector plates are both connected to earth, so there is no potential difference between them. This means that there will be no electric field between them, so there will be no force on the electrons; they will continue at constant velocity in a straight horizontal line. **(2)**

86. Cyclotrons

1 (a) The magnetic field exerts a force on the electron $F = Bev$ at right angles to its velocity. This acts as a centripetal force, so the path is an arc of a circle. The dee is semicircular, so while it is within the dee the electron describes a semicircle. **(2)**

(b) $F = Bev = \frac{mv^2}{r}$; rearrange to give $r = \frac{mv}{Be}$. **(2)**

(c) time = $\frac{\text{distance}}{\text{speed}}$, so $T = \frac{1}{2} \times \frac{2\pi r}{v} = \frac{\pi r}{v}$, but $v = \frac{Ber}{m}$,
so $T = \frac{\pi m}{Be}$, which is independent of v and thus of KE. **(3)**

(d) When the electron travels from left to right, the right-hand dee must be positive and the left-hand dee negative, in order to accelerate the electron between them. When it travels from right to left the polarity must be reversed, so that the electron is always accelerated when it moves from one dee to the other.
The time of orbit remains the same regardless of electron speed or energy, so the polarity has to switch at the same constant frequency that the electron orbits. **(3)**

(e) As their kinetic energy and speed increases the electrons will move into orbits of larger radius, always taking the same time to complete a semicircle. Eventually they will reach the outside of the cyclotron and leave. **(2)**

87. Particle detectors

1 (a) Time for a semicircular circuit $T = \frac{\pi r}{v} = \frac{\pi m}{Be}$, so for a full circuit $\frac{1}{2T} = f = \frac{Be}{2\pi m} = 12.3$ GHz **(3)**

(b) 3.15×10^{-10} J (1.97 GeV) **(3)**

2 (a) Radiation detectors detect ionisation. Neutrons and neutrinos are uncharged, so they do not ionise the air or low-pressure gas inside the detector. **(2)**

(b) (i) The ability to ionise the low-pressure gas. **(1)**
(ii) Alpha particles are strongly ionising but have a very short range. If the end window were thicker, the alpha particles would not penetrate, so no counts would be recorded. **(2)**

88. Matter and antimatter

1 B **(1)**

2 (a) antiproton and antielectron (positron). **(2)**
(b) Mass of a hydrogen atom is approximately 1.67×10^{-27} kg. (The electron mass is negligible compared with the proton mass, but remember to include the mass of the antihydrogen in your calculation too.)
$E = mc^2 = 2 \times (1.67 \times 10^{-27}) \times (3.00 \times 10^8)^2$
$= 3.01 \times 10^{-10}$ J **(4)**
(c) $E = mc^2$
4.2×10^{15} J $= m \times (3.00 \times 10^8)^2$
$m = 0.047$ kg **(2)**

3 (a) pair creation: $^{0}_{0}\gamma \rightarrow \, ^{0}_{-1}e + \, ^{0}_{+1}e$
pair annihilation: $^{0}_{-1}e + \, ^{0}_{+1}e + 2\,^{0}_{0}\gamma$ **(4)**
(b) (i) For one photon (half the total energy),
$E = mc^2 = hf$, so $f = \frac{mc^2}{h}$
wavelength $\lambda = \frac{c}{f} = \frac{h}{mc}$
$$\lambda = \frac{6.63 \times 10^{-34}}{9.11 \times 10^{-31} \times 3.00 \times 10^8}$$
$= 2.43 \times 10^{-12}$ m **(3)**
(ii) If the electron and positron had some KE before they annihilated, this energy would be transferred to the photons along with the energy of the rest masses, so the photons would be more energetic. If so, they would have higher frequency ($E = hf$) and shorter wavelength. **(2)**

89. The structure of nucleons

1 (a) The nuclei of hydrogen atoms are single protons, so the measurement gives their size rather than the size of a nucleus containing several nucleons. Liquid hydrogen is much denser than gaseous hydrogen, so in the liquid there will be more collision events in which electrons scatter from protons. **(2)**

(b) (i) In order to resolve detail, the de Broglie wavelength λ of the electron must be similar to the scale to be measured, the size of the proton. Since $\lambda = \frac{h}{mv}$, for λ to be very small, mv must be very large. If the momentum is large then so is the kinetic energy. **(2)**

(ii) $\lambda = \frac{h}{p} = \frac{6.63 \times 10^{-34}}{2.1 \times 10^{-18}} = 3.2 \times 10^{-16}$ m (0.32 fm) **(2)**

2 (a) Δ_1 (uuu): charge +2
Δ_2 (ddd): charge -1 **(2)**

(b) charge -1 **(1)**

90. Nuclear energy units

1 (a) One electronvolt is the energy transferred when a charge of 1.60×10^{-19} C is accelerated through a potential difference of 1.0 V. It is equal to 1.60×10^{-19} J. **(2)**

(b) 14×10^{12} eV $= 14 \times 10^{12} \times 1.60 \times 10^{-19} = 2.24 \times 10^{-6}$ J (2.24 μJ) **(2)**

(c) KE $= \frac{1}{2}mv^2 = 2.24 \times 10^{-6}$ J, so $\frac{2.24 \times 10^{-6} \times 2}{0.005 \times 10^{-3}} = v^2$
$v = \sqrt{(0.896)} = 0.95$ m s^{-1} **(2)**

2 Units are $\frac{\text{energy}}{\text{speed}^2} = $ J m^{-2} s$^2 = $ kg m^2 s^{-2} m^{-2} s$^2 = $ kg **(2)**

3 (a) rest mass in joules $= 9.11 \times 10^{-31} \times (3.00 \times 10^8)^2$ [kg m^2 s^{-2} = J]
in eV $= \frac{9.11 \times 10^{-31} \times (3.00 \times 10^8)^2}{1.60 \times 10^{-19}} = 5.12 \times 10^5 \frac{\text{eV}}{c^2}$,
so $0.512 \frac{\text{MeV}}{c^2}$ **(1)**

(b) ratio is $\frac{106}{0.512} = 207$ **(1)**

(c) rest mass is $0.512 \frac{\text{MeV}}{c^2}$, so rest energy of e is 0.512 MeV
rest mass is 9.11×10^{-31} kg,
so rest energy is $9.11 \times 10^{-31} \times c^2 = 9.11 \times 10^{-31} \times (3.00 \times 10^8)^2 = 8.20 \times 10^{14}$ J **(2)**

91. The Standard Model

1 (a) Lepton number is conserved. The electron has lepton number +1 and the antineutrino has lepton number -1, so: $0 = 1 - 1$. **(2)**

(b) Z has increased by 1 (one more proton)
$N = A - Z$ has decreased by one (one fewer neutron). **(2)**

(c) (i) The proton and neutron are baryons; the electron and antineutrino are leptons. **(4)**

(ii) 6p + 8n in carbon-14: 14 baryons. 7p and 7n in nitrogen-14: also 14 baryons. **(1)**

2 (a) mesons **(1)**

(b) zero **(1)**

(c) π^0 particle: up and anti-up or down and anti-down

(d) π^+ particle: up and anti-down **(4)**

92. Particle interactions

1 0, 1, 1 **(3)**

2 0, 1, 0 **(3)**

3 Not possible. Charge is conserved but lepton number and baryon number are not conserved. **(3)**

4 (a) mass deficit: $\Delta m = 1.2 - (1 + 0.14)m = 0.06m$.
Therefore energy released $= 0.06mc^2$. **(2)**

(b) $mv_p = 0.14mv_\pi$, so $v_\pi = \frac{1}{0.14} \times v_p \approx 7v_p$ **(3)**

93. Exam skills 10 Nuclear and particle physics

1 (a) The magnetic force is always perpendicular to the particle's velocity, so it provides a centripetal force (Fleming's left-hand rule) and the particle follows the arc of a circle.
$F = Bqv = \frac{mv^2}{r}$
$r = \frac{mv}{Bq}$ **(4)**

(b) (i) Time T taken for one circuit (both dees) $= \frac{2\pi r}{v}$,
so $f = \frac{1}{T} = \frac{v}{2\pi r} = \frac{Bqv}{2\pi mv}$
$f = \frac{Bq}{2\pi m}$ – velocity term cancels, so frequency is independent of velocity **(4)**

(ii) KE of proton is $\frac{1}{2}mv^2 = \frac{1}{2} \times \left(\frac{mBqr}{m}\right)^2$
when $r = \frac{D}{2}$, KE is maximum $= \frac{1}{2} \times \left(\frac{BqD}{2}\right)^2 \times \frac{1}{m}$
$= \frac{B^2q^2D^2}{8m}$ **(4)**

(iii) $E_{\max} = \frac{B^2q^2D^2}{8m} = \frac{(0.8 \times 1.6 \times 10^{-19} \times 1.2)^2}{8 \times 1.67 \times 10^{-27}}$
$= 1.8 \times 10^{-12}$ J
or 11 MeV **(3)**

94. Specific heat capacity

1 $\Delta E = mc\Delta\theta = 2.5 \times 10 \times 2100 = 52\,500$ J **(2)**

2

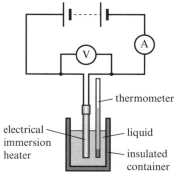

For example: using beaker of liquid and an electric immersion heater. Measure the mass of liquid used using a top-pan balance. Measure the current, voltage and time of heating using an ammeter, voltmeter and stopclock. Measure the temperature rise $\Delta\theta$ using a thermometer. Calculate the energy supplied using $E = IVt$. Calculate the specific heat capacity using $c = \frac{E}{m\Delta\theta}$.

Precautions: lag the beaker and use a lid; stir the liquid to ensure uniform temperature; start timing after heater has heated itself; keep temperature rise low to reduce heat loss as this increases uncertainty in temperature difference – start below and finish above room temperature by same amount. Make sure diagram is fully labelled! **(8)**

3 energy transferred from copper to water $E = m_{\text{Cu}}c_{\text{Cu}}\Delta\theta_{\text{Cu}}$
$= m_{\text{w}}c_{\text{w}}\Delta\theta_{\text{w}}$
$0.250 \times 385 \times \Delta\theta_{\text{Cu}} = 2.0 \times 4200 \times \Delta\theta_{\text{w}}$
$\Delta\theta_{\text{Cu}} = 87.27 \times \Delta\theta_{\text{w}}$
if final temperature of water and copper $= T$,
then $\Delta\theta_{\text{Cu}} = 50 - T$ and $\Delta\theta_{\text{w}} = T - 20$
substitute into first equation:
$50 - T = 87.27 \times (T - 20)$
$50 + 1745.4 = 88.27T$
$\frac{1795.4}{88.27} = T$
$= 20.3°$C. **(4)**

95. Latent heats

1 (a) When the ice melts, it absorbs energy from the drink (the latent heat of fusion). The water formed by the melted ice is at 0°C and absorbs more energy from the drink, cooling it until all the liquid is at the same temperature. If you add water at 0°C, you only get the second cooling effect (which depends only on specific heat capacity and not on latent heat). **(3)**

(b) (i) $E = mL = 50 \times 10^{-3} \times 334 \times 10^3 = 16\,700$ J **(2)**

(ii) $\Delta E = mc\Delta\theta$
$16\,700 = 0.500 \times 4200 \times \Delta\theta$
$\Delta\theta = 8.0$ °C **(2)**

2 (a) • Electrical energy supplied to the heater changes the state of the liquid.

- The amount of energy supplied can be calculated from $E = IVt$ using an ammeter, voltmeter and stopclock.
- The vapour escapes through the holes in the flask and is condensed by the condenser and collected in the beaker.
- This is weighed on a top-pan balance to find the mass of liquid that has changed state.
- The equation to use is: $E = IVt = mL$, where L is the latent heat of vaporisation, $L = \frac{IVt}{m}$ **(6)**

(b) If timing begins before the liquid reaches its boiling point then some of the energy that goes to raising the temperature of the liquid rather than changing its state will be measured. This will result in an overestimate of the latent heat of vaporisation. **(2)**

96. Pressure and volume of an ideal gas

1 (a) In this experiment, the volume of the air is the dependent variable. We measure the length of the air column, which will be directly proportional to the volume as long as the cross-sectional area is constant ($V = \pi r^2 l$). **(2)**

(b) Temperature is a control variable in this experiment. If the gas is compressed quickly, the temperature will rise, causing an additional increase in pressure and Boyle's law will not be followed. **(2)**

(c) A graph of pressure against $\frac{1}{l}$ (length l is proportional to V) will be a straight line through the origin if pressure is inversely proportional to volume. **(4)**

2 (a) $p_1 V_1 = p_2 V_2$

$1.2 \times 10^7 \times 0.0040 = 1.0 \times 10^5 \times V_2$

$V_2 = \frac{1.2 \times 10^7 \times 0.0040}{1.0 \times 10^5} = 0.48 \, \text{m}^3$ **(2)**

(b) Oxygen acts as an ideal gas and obeys Boyle's law. There is no change in temperature. **(2)**

97. Absolute zero

1 D **(1)**

2 Volume of a gas (at constant pressure) extrapolates to zero at about $-273\,°\text{C}$.
Pressure of a gas (at constant volume) extrapolates to zero at about $-273\,°\text{C}$. **(2)**

3 ΔT of $60\,°\text{C}$ (60 K) gives Δp of 23 kPa

Assuming linearity, 1 kPa reduction for each $\frac{60}{23}\,°\text{C}$.

So pressure will be zero at $20 - \left(102 \times \frac{60}{23}\right) = -246\,°\text{C}$. **(3)**

4 (a) 90 K **(1)**
(b) 462 °C **(1)**

5 A real gas would condense to a liquid at some point, so the pressure would drop rapidly, and then change very little. **(3)**

98. Kinetic theory

1 (a) Pressure is force per unit area. Collisions between the container walls and the molecules result in a change of momentum of the molecules. An outward force equal but opposite to the average rate of change of momentum of the molecules is exerted on the walls. **(3)**

(b) When the gas is compressed into a smaller volume, the molecules, which still have the same mean and r.m.s. speed, collide more frequently with the walls, so the rate of change of linear momentum increases and the force increases too. So pressure increases. **(3)**

(c) Heating the gas raises its temperature, so the r.m.s. speed increases. Molecules collide with the walls more frequently and more violently. Both changes increase the rate of change of momentum and hence the force. **(3)**

2 (a) $pV = NkT$

$\frac{110 \times 10^3 \times 3.0 \times 10^{-3}}{298 \times 1.38 \times 10^{-23}} = N$

$= 8.02 \times 10^{22}$ molecules. **(3)**

(b) For one molecule, $\frac{1}{2}m<c^2> = \frac{3}{2}kT$

$= \frac{3}{2} \times 1.38 \times 10^{-23} \times 298 = 6.17 \times 10^{-21}$

multiply by N: $8.02 \times 10^{22} \times 6.17 \times 10^{-21} = 495$ J **(3)**

3 Tiny visible smoke particles are continually bombarded by air molecules. The bombardment is random and changes from moment to moment. The imbalance produces a changing impulse on the smoke particles. **(3)**

99. Particles and energy

1 C **(1)**

2 (a) $441\,\text{m s}^{-1}$ **(1)**
(b) $195662\,\text{m}^2\,\text{s}^{-2}$ **(1)**
(c) $442\,\text{m s}^{-1}$ **(1)**

3 (a) There is a range of kinetic energies. Only a small fraction of the molecules have a mean KE much greater or much lower than the mean KE. **(3)**

(b) T_2: at higher temperatures, more molecules will have greater KE. **(2)**

4 (a) mean KE $= \frac{3}{2}kT = \frac{3}{2} \times 1.38 \times 10^{-23} \times 313$
$= 6.48 \times 10^{-21}$ J **(2)**

(b) mean KE $= \frac{1}{2}m<c^2>$

$<c^2> = \frac{2 \times 6.48 \times 10^{-21}}{4.7 \times 10^{-26}} = 2.75 \times 10^5$

$<c> = 525\,\text{m s}^{-1}$ **(2)**

100. Black body radiation

1 (a) About 0.105 cm. **(1)**

(b) $T = \frac{2.898 \times 10^{-3}}{\lambda_{max}}$ **(2)**

λ_{max} is about 1.05×10^{-3} m, so T is about 2.8 K

(c) Microwave **(1)**

2 (a) Both curves are the shape of black-body radiation curves. The curve for Sirius is always above the curve for Betelgeuse and has a peak at a lower wavelength. The peak wavelength for Sirius is about 3×10^{-7} m. The peak wavelength for Betelgeuse is about 8×10^{-7} m. **(5)**

(b) $L = 4\pi r^2 \sigma T^4 = 4\pi \times (8.2 \times 10^{11})^2 \times 5.67 \times 10^{-8} \times (3500)^4$
$= 7.2 \times 10^{31}$ W **(2)**

101. Standard candles

1 C **(1)**

2 (a) High luminosity means that the stars can be detected at great distances, so this extends the measured distance scale for astronomers. **(2)**

(b) It is only possible to work out the distance of a standard candle if we know both its intensity as viewed from the Earth and its absolute intensity. Distance can then be calculated from the inverse-square law. **(2)**

(c) If the intensity at the Earth is I and the absolute luminosity of the standard candle is L, then
$I = \frac{L}{4\pi d^2}$,
so if we know L and I we can calculate d. **(3)**

3 (a) $I = \frac{L}{4\pi d^2}$

$= \frac{3.8 \times 10^{26}}{4 \times \pi \times (1.5 \times 10^{11})^2} = 1340\,\text{W m}^{-2}$ **(2)**

(b) Power $= 35 \times 1340 \times 0.20 = 9410$ W **(3)**

102. Trigonometric parallax

1 (a) The parallax angles were too small to detect with the naked eye, and telescopes had not been invented. **(1)**

(b) The farther the distance from Earth, the smaller the parallax angle ($\tan \alpha = \frac{r}{d}$, where r = radius of the Earth's orbit and d is the distance to the star). **(1)**

(c) For uniform density of stars, number will be proportional to volume. Volume depends on distance cubed so Hipparcos can measure $10^3 = 1000$ times more stars than ground-based instruments. **(3)**

2 (a) $\tan \alpha = \frac{1.5 \times 10^{11}}{4.24 \times 3.00 \times 10^8 \times 3600 \times 24 \times 365}$

$\alpha = 2.14 \times 10^{-4}$ degrees (0.77 arcseconds) **(3)**

(b) The larger the baseline, the larger the parallax angle and the easier it will be to measure. **(2)**

103. The Hertzsprung–Russell diagram

1 (a) graph with labelled axes (stellar luminosity on y-axis, surface temperature from high to low on x-axis); main sequence, white dwarfs, red giants, red supergiants. **(6)**
(b) middle of main sequence. **(1)**
(c) fusion of hydrogen nuclei into helium nuclei. **(1)**

2 (a) Stellar luminosity: total power radiated by a star. **(2)**
(b) White dwarf star: end stage of the stellar life cycle for a star like our Sun. When fuel runs out, the core collapses under its own gravitational field but stops when it has a very high density. The surface temperature is very high. **(2)**
(c) Neutron star: end stage of the stellar life cycle for a star more massive than our Sun. When fuel runs out, the core collapses under its own gravitational field and atoms are crushed to form a dense ball of neutrons. **(2)**
(d) Black hole: end stage of the stellar life cycle for a star much more massive than our Sun. When fuel runs out, the core collapses under its own gravitational field and the collapse continues without limit. The resulting object is called a black hole because light cannot escape from its gravitational field. **(2)**
(e) Main-sequence star: Once a star has formed and nuclear fusion reactions are taking place in its core, it can remain like this for billions of years. During this stage it is said to be on the main sequence. When main-sequence stars are plotted on the Hertzsprung–Russell diagram, they form a diagonal band. **(2)**

104. Stellar life cycles and the Hertzsprung–Russell diagram

1 (a) mass **(1)**
(b)

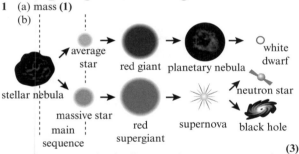

average star → red giant → planetary nebula → white dwarf

stellar nebula

massive star main sequence → red supergiant → supernova → neutron star / black hole **(3)**

2 (a) Stage 1: gas cloud collapses to form a protostar and nuclear fusion reactions begin in the core.
Stage 2: star reaches equilibrium and remains on the main sequence for a long time.
Stage 3: nuclear fuel begins to run out, star begins to collapse, but then restarts fusion and expands to become a red giant.
Stage 4: outer layers of red giant drift away, leaving a white-hot core, a white dwarf star, which gradually cools down. **(6)**
(b) The main difference is in the events at the end of the star's life. Stars much more massive than the Sun swell up to become red supergiants and then undergo a supernova explosion. The remaining core collapses to become either a neutron star or a black hole, depending on its mass. **(4)**

105. The Doppler effect

1 (a) As the car moves from A to C, the component of its velocity directed toward or away from the spectator changes. This causes a Doppler shift, affecting the received frequency. **(2)**
(b) The driver is not moving relative to the engine so there will be no change in the frequency of the sound. **(2)**
(c) (i) From A to B, the frequency heard by the spectator is above the dotted line and falling at increasing rate. At B, the frequency heard by the spectator is equal to that heard by the driver. From B to C, the frequency heard by the spectator is below the dotted line and falling at decreasing rate. **(3)**
(ii) From A to B, there is a component of velocity toward the spectator, so the Doppler shift increases the received frequency.

At B, there is momentarily no component of velocity directed toward the spectator, so the frequency is not Doppler shifted.
From B to C, the component of velocity is away from the spectator, so the Doppler shift reduces the frequency. **(3)**

2 The wavelength is redshifted, so the stars are moving away from the Earth.

Doppler shift $z = \dfrac{\Delta \lambda}{\lambda} = \dfrac{v}{c}$, so speed of recession

$= \dfrac{3.00 \times 10^8 \times (21.132 - 21.106)}{21.106} = 3.7 \times 10^5 \, \text{m s}^{-1}$. **(2)**

106. Cosmology

1 (a) Redshift, $z = \dfrac{\Delta \lambda}{\lambda_0}$, is related to recession velocity v by the equation $z = \dfrac{v}{c}$.

Recession velocity is related to distance by Hubble's law, $v = H_0 d$, so we can combine these two equations to give $zc = H_0 d$ and rearrange this to obtain $d = \dfrac{zc}{H_0}$.

The additional information required is the speed of light and the Hubble constant. **(4)**
(b) The redshift would be affected by any additional (local) motion of the galaxy on top of cosmological expansion. The value of the Hubble constant has some uncertainty. **(2)**

2 (a) If there were a hot Big Bang, then the early Universe would have been filled with high-energy gamma radiation. As the Universe expanded, the radiation would have been redshifted to microwave wavelength characteristic of a low temperature, as it is. **(3)**
(b) The redshifts of other galaxies shows that they are all moving apart. This means that the Universe is expanding. If it is expanding now, it must have started in a much smaller, denser state in the past. **(3)**

3 (a) Hubble's law is $v = H_0 d$, where v is the recession velocity of a galaxy at distance d and H_0 is the Hubble constant. **(3)**

(b) The Hubble time is $\dfrac{1}{H_0} = 13.7$ billion years. When the age is converted to seconds, $H_0 = 2.3 \times 10^{-18} \, \text{s}^{-1}$. **(2)**

107. Exam skills 11 Thermal physics

1 (a) (i) apparatus: sealed flask of gas, heat source, thermometer, pressure gauge **(3)**
(ii) measurements: temperature, pressure **(5)**
(b) Plot pressure against temperature in degrees Celsius. Extrapolate back to intercept on temperature axis. **(2)**
(c) Temperature is related to mean molecular kinetic energy. When the particles have zero kinetic energy that must be the lowest possible temperature. **(2)**
(d) (i) It will be the same, because they are at the same temperature and mean kinetic energy $= \frac{3}{2} kT$. **(2)**
(ii) $\frac{1}{2} m \langle c^2 \rangle = \frac{3}{2} kT$

$\langle c^2 \rangle = \dfrac{3kT}{m} = \dfrac{3 \times 1.38 \times 10^{-23} \times (22 + 273)}{2.7 \times 10^{-25}}$

$= 45\,000$, so $c_{\text{rms}} = 210 \, \text{m s}^{-1}$ **(3)**

108. Mass and energy

1 (a) (i) $m = \dfrac{E}{c^2} = \dfrac{1.43 \times 10^8}{(3.00 \times 10^8)^2} = 1.6 \times 10^{-9} \, \text{kg}$ **(2)**
(ii) This is about one billionth of the mass of the reacting particles and so is negligible. **(2)**
(b) (i) energy per kilogram helium produced E is 6.8×10^{14} J equivalent mass change per kilogram helium produced is $m = \dfrac{E}{c^2} = \dfrac{6.8 \times 10^{14}}{(3.00 \times 10^8)^2} = 7.56 \times 10^{-3} \, \text{kg}$

percentage change $= 0.8\%$ **(2)**
(ii) A change in mass of 0.8% is measurable and not small enough to neglect. **(1)**

(c) Matter–antimatter annihilation is 100% efficient, so:

(i) $\dfrac{1}{1.6 \times 10^{-9}} = 6.3 \times 10^8$ times more efficient than a chemical reaction **(2)**

(ii) $\dfrac{1}{0.008} = 130$ times more efficient than nuclear fusion. **(2)**

2 The total mass annihilated is the sum of the masses of the two particles = $2m_e = 2 \times 9.1 \times 10^{-31}$ kg.
The energy released is given by the equation $E = mc^2$
= $2 \times 9.1 \times 10^{-31} \times (3.00 \times 10^8)^2 = 1.6 \times 10^{-13}$ J **(2)**

109. Nuclear binding energy

1 C **(1)**

2 mass of 8 protons and 8 neutrons = 16.127528 u;
mass deficit = 0.132613 u;
total binding energy = 1.99×10^{-11} J = 125 MeV;
binding energy per nucleon = 7.79 MeV/nucleon
= 1.25×10^{-12} J/nucleon. **(4)**

3 (a) Iron-56 has the greatest value of binding energy per nucleon, so it takes more energy per nucleon to split the nucleus up into individual nucleons than any other nucleus. **(2)**

(b) All three of these nuclides are peaks on the binding energy per nucleon curve, so they are relatively more stable than the nuclides adjacent to them. This means that, when nuclei form, these will be more likely to form and so are produced in larger numbers (by nuclear fusion reactions in stars). **(3)**

110. Nuclear fission

1 (a) Main features: axes labelled, binding energy per nucleon on y-axis and nucleon or mass number on x-axis; steep rise to peak at iron-56; shallow drop to uranium at end (see page 109). **(3)**

(b) Splitting a heavy nucleus (e.g. uranium) can form two lighter nuclei. These have greater binding energy per nucleon, so the energy difference can be released in the nuclear fission process. **(3)**

2 $a = 1$; $b = 0$; $c = 141$; $d = 56$. **(4)**

3 (a) 200 MeV per fission = $200 \times 10^6 \times 1.6 \times 10^{-19}$
= 3.20×10^{-11} J
mass equivalent $m = \dfrac{E}{c^2} = \dfrac{3.20 \times 10^{-11}}{(3.00 \times 10^8)^2} = 3.56 \times 10^{-28}$ kg
= 0.213 u **(3)**

(b) $N = \dfrac{1.0}{235} \times 6.02 \times 10^{23}$ fissions
Total energy = $3.20 \times 10^{-11} \times \dfrac{1.0}{235} \times 6.02 \times 10^{23}$
= 8.2×10^{10} J **(3)**

111. Nuclear fusion

1 (a) Both involve nuclear transformations and release nuclear energy.
Fission involves the splitting of heavy nuclei; fusion involves the joining together of light nuclei. **(2)**

(b) Two nuclei have to approach one another close enough for the strong nuclear force to overcome the coulomb (electrostatic) repulsion, since both nuclei are positively charged. The strong nuclear force can then bind them together to form a heavier nucleus with the release of energy. This is very difficult to achieve in practice because the individual nuclei must have extremely high energies to get so close. This means creating and controlling plasmas at extremely high temperatures. **(4)**

(c) (i) Charge: the proton numbers (the lower ones) represent the charges on the nuclei, and the sum is the same on each side (1 + 1 = 1 + 1)
Baryon number: the baryon numbers (numbers of nucleons) on the top are also equal on each side (2 + 2 = 3 + 1) **(2)**

(ii) mass deficit = $2 \times 2.013553 - (1.007276 + 3.015500)$
= 4.33×10^{-3} u = 7.23×10^{-30} kg
energy = $mc^2 = 7.23 \times 10^{-30} \times (3.00 \times 10^8)^2$
= 6.51×10^{-13} J, 4.07 MeV **(5)**

(iii) number of hydrogen and deuterium nuclei in 1.0 kg
water = $2 \times \dfrac{1.0}{0.018} \times 6.02 \times 10^{23}$
number of deuterium nuclei only
= $\dfrac{2 \times \dfrac{1.0}{0.018} \times 6.02 \times 10^{23}}{4500} = 1.49 \times 10^{22}$
maximum energy released
= $1.49 \times 10^{22} \times 6.51 \times 10^{-13} = 9.68 \times 10^9$ J **(3)**

112. Background radiation

1 C **(1)**

2 (a) Nuclear explosions involve fission reactions. These produce radioactive daughter nuclei. **(2)**

(b) Once the test ban treaty was in place, nations stopped testing weapons in the atmosphere, so there was no new contamination. The existing contamination decayed and eventually became stable. After 50+ years, only isotopes with long half-lives remain. **(2)**

3 Each window has a different absorber: none, aluminium, lead.
The open window allows alpha, beta and gamma radiation to reach the film behind it.
The thin aluminium window can only be penetrated by beta and gamma.
The lead window lets only gamma through.
By comparing these, it is possible to work out how much of each type of radiation has been absorbed by the person wearing the badge. **(4)**

113. Alpha, beta and gamma radiation

1 (a) Gamma rays are uncharged, so they do not interact so strongly with matter. This means that they lose energy more slowly and travel farther than alpha or beta particles. **(2)**

(b) Alpha and beta particles are both charged, so they experience electrostatic forces in an electric field and, since they are moving, magnetic forces in a magnetic field. Gamma rays, being uncharged, do not. **(2)**

(c) Alpha particles are strongly ionising but have very short range. They can be stopped by the outer layers of (mainly) dead skin. However, if they are ingested, they can cause great damage to growing and dividing cells. **(2)**

2 (a) Radioactivity is a random process. **(1)**

(b) 81.2, 75.8 **(2)**

(c) Beta particles. There is a small reduction when thick card is used. If the source were an alpha source, this would have fallen back to background levels. Beta and gamma radiation can penetrate this (although a few beta particles would be absorbed). When lead is used, the count rate falls back to background levels, suggesting that all the radiation from the source has been absorbed. If it were a gamma emitter, these readings would be higher. **(4)**

(d) Handle the source by using tongs; point source away from body; shield source and experiment from student; limit time of use; maximise distances to source; wash hands after doing experiment; don't eat or drink in lab; ... **(3)**

114. Investigating the absorption of gamma radiation by lead

1 B **(1)**

2 (a) m^{-1} **(1)**

(b) (i) micrometer screw gauge or Vernier callipers **(1)**

(ii) Remove source. Measure counts over, say, 5 minutes. Repeat at least three times. Find the average number of counts per minute. **(2)**

(iii) $\ln(I) = \ln(I_0) - \mu x$. This has the form $y = mx + c$. y is $\ln(I)$. m is $-\mu$. The graph of $\ln(I)$ versus x will be a straight line with a negative gradient equal to $-\mu$. **(4)**

115. Nuclear transformation equations

1 (a) charge: $0 = 1 - 1$; conserved
baryon number: $1 = 1 + 0$; conserved **(2)**

(b) lepton number: $0 \neq 0 + 1$; not conserved **(1)**

(c) $^1_0n \rightarrow {}^1_1p + {}^0_{-1}e + {}^0_0\overline{\upsilon}$

The antineutrino has lepton number -1, so it cancels the $+1$ for the electron created in the decay and lepton number is conserved. **(3)**

2 (a) (i) $^{234}_{90}Th \rightarrow {}^{234}_{91}Pa + {}^0_{-1}e + {}^0_0\overline{\upsilon}$ **(3)**

(ii) $^{230}_{90}Th \rightarrow {}^{226}_{88}Ra + {}^4_2\alpha$ **(3)**

(b) (i) Change in baryon number is $232 - 208 = 24$. Therefore 6 alpha particles emitted. **(1)**

(ii) 12 protons lost in alpha particles, but change in proton number is $90 - 82 = 8$ protons. Therefore 4 neutrons in nucleus decay to protons, therefore 4 beta particles emitted. **(1)**

116. Radioactive decay and half-life

1 A **(1)**

2 (a) 215, 159, 116, 88, 69, 54, 45, 38, 36 **(2)**

(b)
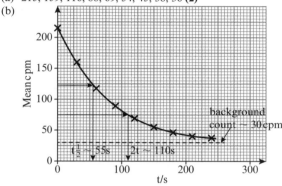

(c) About 30–35 cpm (judged from flattening of graph). **(1)**

(d) About 55 s. Working should show an attempt to find the half-life from more than one starting point. **(3)**

117. Exponential decay

1 (a) 3.71, 3.26, 2.83, 2.40, 1.79 **(2)**

(b)
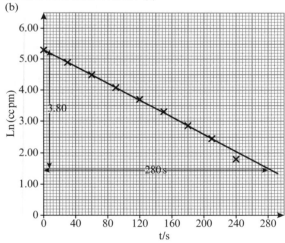

Decay constant λ is negative gradient of graph.

Gradient $= -\dfrac{3.80}{280} = -0.0136\,s^{-1}$

$\lambda = 0.0136\,s^{-1}$

$t_{1\backslash2} = \dfrac{\ln 2}{\lambda}$. Half-life = 50 s (approximately) **(6)**

118. Radioactive decay calculations

1 (a) The gamma rays are emitted randomly in all directions, so the detector only records a fraction of them.

The efficiency of the detector is not 100% − the detector only records a fraction of the gamma rays that pass through it.

The body of the sample itself might absorb some of the gamma rays inside it. **(3)**

(b) (i) After one half-life, $A = \dfrac{186}{2} = 93\,kBq$ **(1)**

(ii) After three half-lives, $A = \dfrac{186}{2^3} = 23\,kBq$ **(2)**

(c) (i) $A = A_0 e^{-\lambda t}$, so after one half-life ($t = 5.3$ years)

$1 = 2e^{-\lambda t}$

$\ln 0.5 = -\lambda t_{1\backslash2}$

$\lambda = -\dfrac{\ln 0.5}{5.3 \times 365 \times 24 \times 60 \times 60}$

$= 4.1 \times 10^{-9}\,s^{-1}$ **(1)**

(ii) $A = A_0 e^{-\lambda t}$, so when $A = \dfrac{A_0}{10}$, then $\dfrac{1}{10} = e^{-\lambda t}$

Taking natural logs, $\ln 0.1 = \ln(e^{-\lambda t}) = -\lambda t$

$-\dfrac{2.30}{4.1 \times 10^{-9}} = -t$

$t = 5.56 \times 10^8\,s$, so every 17.6 years. **(3)**

119. Gravitational fields

1 C **(1)**

2 The time of flight (time to fall to the ground) increases by a factor of $\dfrac{9.81}{1.63}$, or 6.0 times. The horizontal component of the velocity is the same as on Earth, so the range will be at least 6 times greater (1500 m). In addition to this, there is no atmosphere on the Moon, so the ball's average speed will be greater. It can travel considerably farther than 1500 m and so may well have a range in excess of a mile (1600 m). **(4)**

3 B **(1)**

4 (a) constant in magnitude and direction at all points. **(1)**

(b) Over distances small compared with the radius of the Earth, the angle between field lines is so small that they can be treated as parallel and equally spaced. **(1)**

120. Gravitational potential and gravitational potential energy

1 (a) $\Delta V_g = g\Delta h = 9.81 \times 560 \times 10^3 = 5.49 \times 10^6\,J\,kg^{-1}$ **(2)**

(b) Smaller, because g gets smaller as we move farther from the Earth's surface. **(2)**

2 gravitational potential energy of tower (using height h of centre of mass, 16.0 cm) is $mgh = 8 \times 0.200 \times 9.81 \times 0.160$

gravitational potential energy of all bricks on floor (using height h of centres of mass, 2.0 cm) is $mgh = 8 \times 0.200 \times 9.81 \times 0.020$

$\Delta GPE = 8 \times 0.200 \times 9.81 \times (0.160 - 0.020) = 2.20\,J$ **(4)**

121. Newton's law of gravitation

1 (a) $F = \dfrac{Gm_1m_2}{r^2}$, so $\dfrac{F_A}{F_B} = \dfrac{Gm_1m_2}{r_A{}^2} \times \dfrac{r_B{}^2}{Gm_1m_2} = \dfrac{r_B{}^2}{r_A{}^2}$

$= \dfrac{249^2}{207^2} = 1.45$ **(2)**

(b) Toward the Sun. **(1)**

(c) Speed is increasing and the radius of curvature of its path is getting smaller. **(1)**

(d) At B, Mars is at its furthest point from the Sun, so it has its greatest gravitational potential energy. As it accelerates around its orbit, the gravitational potential energy decreases (becomes more negative), transferring to kinetic energy. At A, the planet is closest to the Sun and has its maximum kinetic energy and least gravitational potential energy. As it moves along its orbit back toward B, kinetic energy is transferred back to gravitational potential energy. **(4)**

(e) At A. **(1)**

(f) maximum gravitational force at A (closest point)

$F_A = \dfrac{Gm_1m_2}{r_A{}^2} = \dfrac{6.67 \times 10^{-11} \times 6.39 \times 10^{23} \times 1.99 \times 10^{30}}{(207 \times 10^9)^2}$

$= 1.98 \times 10^{21}\,N$ **(2)**

2 (a) $g = \dfrac{Gm}{r^2} = \dfrac{6.67 \times 10^{-11} \times 7.35 \times 10^{22}}{(1740 \times 10^3)^2} = 1.62\,N\,kg^{-1}$ **(2)**

(b) $W = mg = 65 \times 1.62 = 105\,N$ **(1)**

122. Gravitational field of a point mass

1 (a) $g = \dfrac{Gm}{x^2}$ **(1)**

(b) (i) $g_{neutral} = \dfrac{Gm_E}{a^2} = \dfrac{Gm_M}{b^2}$

$\dfrac{a^2}{b^2} = \dfrac{m_E}{m_M}$

$\dfrac{a}{b} = \sqrt{\dfrac{m_E}{m_M}} = \sqrt{80} = 8.9$ **(3)**

(ii) $\dfrac{a}{(a+b)} = \dfrac{8.9}{9.9}$, and $(a+b) = 3.84 \times 10^8$ m

$a = \dfrac{8.9}{9.9} \times 3.84 \times 10^8 = 3.45 \times 10^8$ m **(2)**

(c) Work must be done to reach the neutral point, but beyond it the spacecraft will 'fall' toward the Moon. **(2)**

2 (a) As you go beneath the surface of the planet, some of the matter closer to the surface (above you) exerts a force on you away from the centre of the planet. This reduces the overall force per unit mass (gravitational field strength) at that point. **(2)**

(b) Zero. By symmetry all of the gravitational forces acting on a mass at the centre must cancel out. **(2)**

123. Gravitational potential in a radial field

1 (a) The gravitational potential at a point in space is the gravitational potential energy per unit mass at that point. **(2)**

(b) Gravitational potential is zero at infinite distance, and all gravitational forces are attractive. This means that work must be done to move any mass from a point in space to infinity. If energy must be supplied to raise the potential to zero, then it must have started as a negative value. **(2)**

(c) Gravitational field strength is the negative gradient of the gravitational potential. Equipotential surfaces are perpendicular to gravitational field lines. **(2)**

(d) The change in GPE is equal to the change in gravitational potential between two points multiplied by the mass moved between those points. **(2)**

2 (a) Equipotentials at 9.81 J kg^{-1}, 19.6 J kg^{-1}, 29.4 J kg^{-1}, etc.

(b) Vertical and horizontal distances are small compared with the radius of the Earth. **(2)**

(c) Radial field lines pointing toward the Earth. Concentric equipotentials with increasing separation as you move away from the Earth. **(4)**

124. Energy changes in a gravitational field

1 (a) vertical height $= 125 \times \sin 5° = 10.9$ m

increase in gravitational potential energy $= mgh$
$= 90 \times 9.81 \times 10.9 = 9620$ J
kinetic energy at bottom of hill $= \frac{1}{2}mv^2 = \frac{1}{2} \times 90 \times (16)^2$
$= 11520$ J
Yes, he can freewheel to the top of the hill. **(4)**

(b) It makes no difference because mass cancels from the equation (both KE and GPE are directly proportional to mass). **(2)**

2 (a) at Earth's surface $V_g = -\dfrac{GM}{r} = -\dfrac{6.67 \times 10^{-11} \times 5.97 \times 10^{24}}{6.37 \times 10^6}$

$= -6.25 \times 10^7$ J kg^{-1}

at altitude of satellite $V_g = -\dfrac{GM}{r} =$

$-\dfrac{6.67 \times 10^{-11} \times 5.97 \times 10^{24}}{(35.8 + 6.37) \times 10^6} = -9.44 \times 10^6$ J kg^{-1} **(4)**

(b) $\Delta E_g = m\Delta V_g = 3000 \times (6.25 \times 10^7 - 0.944 \times 10^7)$
$= 1.59 \times 10^{11}$ J **(2)**

(c) E_g (on the ground) $= mgh = 3000 \times 9.81 \times 6.37 \times 10^6$
$= 1.87 \times 10^{11}$ J
Total E_g (in orbit) $= -1.87 \times 10^{11} + 1.59 \times 10^{11}$
$= -2.8 \times 10^{10}$ J **(2)**

125. Comparing electric and gravitational fields

1 B **(1)**

2 gravitational $F = \dfrac{Gm_1m_2}{r^2}$

$= \dfrac{6.67 \times 10^{-11} \times 6.64 \times 10^{-27} \times 3.27 \times 10^{-25}}{r^2} = \dfrac{1.45 \times 10^{-61}}{r^2}$

electrostatic $F = \dfrac{Q_1Q_2}{4\pi\varepsilon_0 r^2} = \dfrac{3.20 \times 10^{-19} \times 1.27 \times 10^{-17}}{4 \times \pi \times 8.85 \times 10^{-12} \times r^2}$

$= \dfrac{3.65 \times 10^{-26}}{r^2}$

The electrostatic force is more than 10^{35} times greater than the gravitational attraction, so Rutherford was justified in ignoring gravitational effects. **(4)**

3 (a) For example, two positive charges placed close to one another. It would require work to be done to put them into this configuration because they exert repulsive forces on each other. If work is done, their electrostatic potential energy must increase from zero (the electrostatic potential energy at infinity). **(3)**

(b) All gravitational forces are attractive, so the field does work to pull them together from infinity, and work would have to be done to separate them. This means that the gravitational potential energy (GPE) at infinity (zero) is greater than that of any configuration of masses, so they must have a negative GPE. **(2)**

126. Orbits

1 (a) speed of satellite $v = \dfrac{2\pi R}{T}$

centripetal force $=$ gravitational force, $F = \dfrac{mv^2}{R}$

$= \dfrac{GMm}{R^2}$

$\dfrac{m\left(\dfrac{2\pi R}{T}\right)^2}{R} = \dfrac{GMm}{R^2}$

$T = \sqrt{\dfrac{4\pi^2 R^3}{GM}}$, which is independent of m. **(3)**

(b) (i) GPE $= -\dfrac{GMm}{R}$ **(1)**

(ii) KE $= \frac{1}{2}mv^2 = +\frac{1}{2}\dfrac{GMm}{R}$ **(2)**

(iii) $\dfrac{\text{GPE}}{\text{KE}} = -2$ **(1)**

2 (a) KE is maximum at A, closest to Earth, and falls as spacecraft moves from A to B. GPE rises from a minimum at A to a maximum at B. Then GPE falls as it moves from B to A and KE rises from B to A. From A to B, KE is being transferred to GPE and from B to A, GPE is being transferred to KE. The total energy is constant. **(3)**

(b) The departure point will not affect the amount of fuel because the total energy of the spacecraft is constant in the orbit. This means it needs the same additional energy from burning fuel to get to infinity (zero GPE) from any point on its orbit. **(4)**

127. Exam skills 12 Gravitational fields

1 (a) The gravitational force of attraction between two point masses m_1 and m_2 a distance r apart is directly proportional to the product of their masses and inversely proportional to the square of their separation:

$F = \dfrac{Gm_1m_2}{r^2}$. **(1)**

(b) gravitational force per unit mass at a point in space, N kg^{-1}. **(2)**

(c) gravitational force $=$ centripetal force, $F = \dfrac{Gm_1m_2}{r^2}$

$= mr\omega^2$, where $\omega = \dfrac{2\pi}{T}$

orbital radius $r^3 = \dfrac{Gm_E T^2}{4\pi^2}$

$= \dfrac{6.67 \times 10^{-11} \times 6.0 \times 10^{24} \times (2 \times 3600)^2}{4\pi^2} = 5.26 \times 10^{20}$

$r = 8.067 \times 10^6$ m
altitude $= r - 6.400 \times 10^6 = 1.7 \times 10^6$ m (1700 km) **(4)**

(d) (i) Work that must be done per unit mass to move a mass from infinity (zero potential) to the point in space. **(2)**

(ii) Gravitational potential energy (GPE) at infinity is zero. Work must be done against attractive forces to move a mass from a point in space to infinity, so the GPE at the initial point must be less than at infinity, i.e. negative. **(3)**

(e) (i) In 1 year, ΔGPE $= -\dfrac{GMm}{r_2} - -\dfrac{GMm}{r_1}$

$= -6.67 \times 10^{-11} \times 6.0 \times 10^{24} \times 7.3 \times 10^{22}$

$\times \left(\dfrac{r_1}{r_1 r_2} - \dfrac{r_2}{r_1 r_2}\right)$

r_1r_2 is approximately r^2, so ΔGPE

$= -2.92 \times 10^{37} \times \frac{-38 \times 10^{-3}}{(3.9 \times 10^8)^2}$

$= 7.3 \times 10^{18}$ J/year = 230 GW **(3)**

(ii) Moon's KE reduces. Earth's rotational KE reduces. **(2)**

128. Simple harmonic motion

1 Record the time for 10 oscillations. Repeat three times to obtain an average value of $10T$ and divide by 10 to find T. Ignore or repeat anomalous results. Use a fiducial marker placed at the centre of the oscillations to reduce timing errors. Calculate f from $f = \frac{1}{T}$. **(5)**

2 (a) Force is directly proportional to displacement from equilibrium position. Force is directed toward equilibrium position. **(2)**

(b) The force is directed toward the equilibrium position and is directly proportional to displacement for initial displacements up to 9.0 mm. For small initial amplitudes, the conditions for simple harmonic motion (S.H.M.) are met, so it would undergo S.H.M. However, for larger initial displacements, the restoring force is not directly proportional to displacement so the oscillations would not be simple harmonic. **(4)**

(c) A shorter length would be stiffer, so larger forces would be produced for the same deflection. There is also less mass, so the accelerations would be larger $\left(a = \frac{F}{m}\right)$ and time period would be smaller. **(2)**

129. Analysing simple harmonic motion

1 A **(1)**

2 (a) $t = \frac{1}{f} = 0.50$ s **(1)**

(b) (i) $x = A\cos(2\pi ft) = 8.0\cos(2\pi \times 2.0 \times 0.125) = 0$ cm **(1)**

(ii) $x = 8.0\cos(2\pi \times 2.0 \times 0.25) = -8.0$ cm **(1)**

(iii) $x = 8.0\cos(2\pi \times 2.0 \times 0.40) = +2.5$ cm **(1)**

(c) $v = -A2\pi f\sin(2\pi ft)$ so max $= -A2\pi f = 1.01$ m s^{-1} as the oscillator passes its equilibrium position. **(2)**

(d) $F = -kx = -m\omega^2 x$

$= -m(2\pi f)^2 x = 0.25 \times 4\pi^2 \times 2.0^2 \times 0.08 = 3.2$ N at maximum amplitude. **(2)**

(e) Maximum KE (at maximum velocity) $= \frac{1}{2} \times mv^2$

$= \frac{1}{2} \times 0.25 \times (1.01)^2 = 0.13$ J

At this point, the potential energy is 0 (equilibrium position), so total maximum energy = 0.13 J **(2)**

3 $v = -\omega A\sin(2\pi ft)$; $a = -\omega^2 A\cos(2\pi ft)$ **(2)**

130. Graphs of simple harmonic motion

1 (a) 64 mm **(1)**

(b) 3T = 151 ms, so frequency is 19.9 Hz or 20 Hz **(1)**

(c) Positive cosine curve with maximum velocity at 8.0 m s^{-1} **(4)**

(d) Negative sine curve with maximum acceleration at 1000ms^{-2} **(4)**

131. The mass–spring oscillator and the simple pendulum

1 C **(1)**

2 (a) $T = 2\pi\sqrt{\frac{m}{k}} = 0.795$ s **(2)**

(b) The spring itself has mass, and this is also oscillating. This increases the effective mass of the system, and since $T = 2\pi\sqrt{\frac{m}{k}}$ it increases the time period. **(3)**

3 (a) $T = 2\pi\sqrt{\frac{l}{g}}$, so $l = \frac{gT^2}{4\pi^2} = \frac{9.81 \times 1.00^2}{4\pi^2} = 0.248$ m **(2)**

(b) $\frac{\Delta T}{T} = \frac{1}{2}\frac{\delta l}{l} = 0.0025 = 0.25\%$

[Long form calculation:] $\Delta T = 2\pi\sqrt{\frac{l_2}{g}} - 2\pi\sqrt{\frac{l_1}{g}}$,

so percentage change $= \frac{\Delta T}{T} \times 100\%$

$= \frac{(\sqrt{l_2} - \sqrt{l_1})}{\sqrt{l_1}} \times 100\%$

$= (\sqrt{1.005} - 1) \times 100\% = 0.25\%$ **(3)**

132. Energy and damping in simple harmonic oscillators

1 D **(1)**

2 (a) gravitational potential energy, kinetic energy, elastic potential energy: GPE→KE→EPE→KE→GPE **(2)**

(b) $x = -\frac{mg}{k} = -\frac{0.500 \times 9.81}{42} = 0.117$ m **(2)**

(c) (i) $E = \frac{1}{2}kx^2 = \frac{1}{2} \times 42 \times (0.117)^2 = 0.287$ J **(1)**

(ii) $E = \frac{1}{2} \times 42 \times (0.097)^2 = 0.197$ J **(1)**

(iii) $E = \frac{1}{2} \times 42 \times (0.137)^2 = 0.394$ J **(1)**

(d) $\Delta GPE = mg\Delta h = 0.500 \times 9.81 \times 0.04 = 0.197$ J $= 0.394 - 0.197$ J **(2)**

(e) maximum KE = 0.197 J; $v = \sqrt{\frac{2E}{m}} = 0.889$ m s^{-1} **(3)**

133. Forced oscillations and resonance

1 (a) (i) Natural frequency is the frequency of oscillations when the oscillator is displaced and released with no external forces acting on it. Driving frequency is the frequency of the external forces (in this case from the vibration generator) acting on the system. **(2)**

(ii) $f = \frac{1}{2\pi} \times \sqrt{\frac{k}{m}} = \frac{1}{2\pi} \times \sqrt{\frac{30}{0.065}}$

= 3.42 Hz **(1)**

(b) (i) Curve starts at A_0, peaks at about 3.4 Hz, decays asymptotically toward zero. **(3)**

(ii) Resonance occurs when the driver (forcing) frequency is very close to the natural frequency of the driven oscillator. Resonance results in a large amplitude of response. **(2)**

(iii) At resonance, the driver continually transfers energy to the driven oscillator, causing its amplitude to grow. At the same time, work done by the oscillator against damping forces transfers energy to heat. The amplitude at resonance is determined by the point at which these two energy transfers occur at an equal rate. **(4)**

134. Driven oscillators

1 (a) Curve starts at A_0, peaks at 6.59 Hz, decays asymptotically toward zero. (Natural frequency $f = \frac{1}{2\pi} \times \sqrt{\frac{k}{m}} = 6.59$ Hz.) **(4)**

(b) Curve starts at A_0; lower peak at resonance; curve lies below undamped curve; frequency of resonance slightly lower than natural frequency of oscillator. **(2)**

(c) Natural frequency is given by $f = \frac{1}{2\pi} \times \sqrt{\frac{k}{m}}$, so the student could increase k (use stiffer springs) or reduce m (use a smaller mass). **(3)**

135. Exam skills 13 Simple harmonic motion

1 (a) (i) $f = \frac{1}{2\pi}\sqrt{\frac{6500}{75}} = 1.48 \approx 1.5$ Hz **(3)**

(ii) $E = \frac{1}{2}kx^2 = \frac{1}{2} \times 6500 \times 0.084^2 = 23$ J **(2)**

(iii) maximum $a = -A(2\pi f)^2 = -0.084 \times 4\pi^2 \times (1.5)^2 = -7.3$ m s^{-2} **(2)**

(b) Max acceleration = (7.26 + 9.81) so

Max force = ma = 60 × 17.1 = 1025 N ≈ 1030 N **(2)**

(c) (i) If the bicycle continued to undergo S.H.M. after hitting the bump, it would be very difficult to ride. The damper dissipates the energy of the oscillation, so that the bike returns to normal after hitting the bump. **(2)**

(ii) 8.4 cm marked on compression axis at point where graph starts. Two cycles of oscillation with amplitude decaying to zero. **(4)**

Published by Pearson Education Limited, 80 Strand, London, WC2R 0RL.

www.pearsonschoolsandfecolleges.co.uk

Copies of official specifications for all Edexcel qualifications may be found on the website: www.edexcel.com

Text and illustrations © Pearson Education Limited 2016
Typeset and illustrations by Tech-Set Ltd, Gateshead
Produced by Out of House Publishing
Cover illustration by Miriam Sturdee

The rights of Steve Adams and John Balcombe to be identified as authors of this work have been asserted by them in accordance with the Copyright, Designs and Patents Act 1988.

First published 2016

18 17
10 9 8 7 6 5 4 3

British Library Cataloguing in Publication Data
A catalogue record for this book is available from the British Library

ISBN 9781447989950

Printed in the UK by Ashford Colour Press Ltd

A note from the publisher
In order to ensure that this resource offers high-quality support for the associated Pearson qualification, it has been through a review process by the awarding body. This process confirms that this resource fully covers the teaching and learning content of the specification or part of a specification at which it is aimed. It also confirms that it demonstrates an appropriate balance between the development of subject skills, knowledge and understanding, in addition to preparation for assessment.

Endorsement does not cover any guidance on assessment activities or processes (e.g. practice questions or advice on how to answer assessment questions), included in the resource nor does it prescribe any particular approach to the teaching or delivery of a related course.

While the publishers have made every attempt to ensure that advice on the qualification and its assessment is accurate, the official specification and associated assessment guidance materials are the only authoritative source of information and should always be referred to for definitive guidance.

Pearson examiners have not contributed to any sections in this resource relevant to examination papers for which they have responsibility.

Examiners will not use endorsed resources as a source of material for any assessment set by Pearson.

Endorsement of a resource does not mean that the resource is required to achieve this Pearson qualification, nor does it mean that it is the only suitable material available to support the qualification, and any resource lists produced by the awarding body shall include this and other appropriate resources.